389
Dem

Deming, Richard

Metric power

12591

DATE DUE			
DE 2 '77			
MAY 1 1 79			
OC 30 '87			

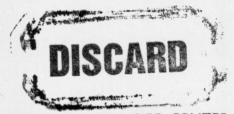

Metric Power

Also by Richard Deming

Heroes of the International Red Cross
Man Against Man: Civil Law at Work
Man and Society: Criminal Law at Work
Man and the World: International
        Law at Work
Police Lab at Work
Sleep, Our Unknown Life
Vice Cop
Vida

METRIC POWER

WHY AND HOW WE ARE GOING METRIC

by Richard Deming

THOMAS NELSON INC., PUBLISHERS

Nashville, Tennessee / New York, New York

Third printing

Library of Congress Cataloging in Publication Data

Deming, Richard.
    Metric power; why and how we are going metric.

    SUMMARY: Discusses the present system of measurements in the United States and the distinct advantage of using the metric system, including its effect on business, industry, and the individual.
    Bibliography: p.
    1. Metric system. 2. Weights and measures—United States.
[1. Metric system. 2. Weights and measures]
I. Title.
QC92.U54D45   389'.152   74–5039
ISBN 0–8407–6380–8

For Otie

Contents

The world hates change, yet it is the only thing that has brought progress.

Charles Franklin Kettering

Metric Power

Chapter I

All Over but the Shouting

At the end of the nineteenth century an essay in an American religious journal read in part:

> This system came out of the Bottomless Pit. At that time and in the place whence this system sprang it was hell on earth. The people defied the God who made them; they worshipped the Goddess of Reason. In their mad fanaticism they brought forth monsters . . . unclean things. Can you, the children of the Pilgrim Fathers, worship at such a shrine, and force upon your brethren the untimely monster of such an age and place?

To what evil cult with unspeakable rituals do you suppose the writer was referring? Satanism, perhaps?

Actually, he was not referring to any cult. He was merely registering an objection to the adoption of the metric system in the United States.

A few years later, in 1904, Frederick A. Hasley, who later became "commissioner" of a privately sponsored antimetric organization that called itself the American Institute of Weights and Measures, published a book titled *The Metric Fallacy* in collaboration with a man named Samuel Dale. Chapter I began with these words:

> The metric system was adopted in France by a compulsory law of the most drastic character in 1793. That law remained in force

13

for nineteen years, or until 1812 when, under Napoleon who had no faith in the system, the law was repealed and the people were permitted to resume the use of their old measures. . . . Under relaxed laws, the French people immediately reverted to that truly universal system in which twelve inches make a foot, three feet make a yard, and sixteen ounces make a pound and this practice continued for twenty-five years, or until 1837 when the metric force laws were reimposed, and they have been continued in force until the present day.

Why did the French people revert to their old system as soon as they were given the opportunity? Are not nineteen years of enforced use sufficient to demonstrate the advantages of the metric system if such advantages exist? What explanation of this experience is possible except that the French people found the old system better than the new for the uses to which a system of weights and measures is applied? . . . Is it not clear that the people everywhere do not like the metric system because they find it inferior to the older system for the purposes of everyday life?

Of course, both of these quotations are from a much less enlightened era than ours. It would be natural to assume that in this more scientific age such passionate opposition to something as philosophically neutral as a measuring system is unlikely.

Natural, perhaps, but not necessarily correct. Consider these two incidents:

On April 19, 1973, Wilson Riles, Superintendent of Public Instruction for the State of California, called a press conference to announce that beginning in 1976 the metric system would be taught exclusively in the California public schools in place of the inch-quart-pound system. For the most part the reaction, both of the news media and of individuals who wrote or wired comments about the announcement to Superintendent Riles, was favorable. But one woman wrote that conversion to metrics was "part of a conspiracy to brainwash the children to favor communism." As evidence in support of her charge, she pointed out that not only the Soviet Union, but every Communist nation in the world uses metrics.

The second incident occurred in September 1973 on the second day of a two-day conference on metrics held at the University of California at Los Angeles and attended by nearly 750 representatives of educational systems, industries, and various levels of government from all over the United States. There appeared on the campus to hand out antimetric pamphlets, and to buttonhole any delegates polite enough to listen, a member of a right-wing organization so bitterly opposed to conversion to metrics that it had invented an entirely new measuring system of its own. According to the pamphlet, copies of the system were available for $100 each.

There is a distinct and important difference between past opposition to metric conversion and that encountered today, however. Until relatively recently a good deal of the opposition was from highly respected and influential sources. Frederick Hasley, for instance, was a graduate engineer with a degree from Cornell University and was associate editor of a technical magazine called the *American Machinist*. Samuel Dale was editor of two textile magazines. Their antimetric organization, the American Institute of Weights and Measures, drew much of its financial support from American industry, which feared the cost of having to convert its machines and tools to metric.

Although there is still some opposition to conversion from highly respected sources, most opposition today is from uninformed persons who simply have no understanding of the issues. For the most part, their objections are based on such emotional vagaries as the belief that the inch-quart-pound system was approved by God or that metrication is some kind of sinister plot to aid a Communist take-over.

Actually, it is all over but the shouting, because the battle has been won. We are already so far along the road to metrication that there is no chance of turning back. Opposition from industry has not entirely disappeared, but a recent survey by the United States National Bureau of Standards showed that 70 percent of American manufacturers favored metrication.

Many industries are already in the process of converting to metrics. Among the larger ones are the Ford Motor Company, General Motors, International Harvester, and International Business Machines.

In the educational field the following national associations are on record as wholeheartedly supporting conversion: the National Education Association, the National Association of Secondary School Principals, the National Council of Teachers of Mathematics, the Council for Exceptional Children, the Association of American Colleges, the Association of Classroom Teachers, the National Science Teachers Association, the American Society for Engineering Education, the Association for Educational Communications and Technology, and the National Congress of Parents and Teachers. Regional and statewide educational associations in support of conversion are too numerous to attempt to list.

Engineering and technological associations and societies are almost unanimously in favor of conversion.

In some areas the United States has already quietly converted to metrics. The most popular film size for still cameras is 35 mm. Movie film comes in 8 mm and 16 mm sizes. The pharmaceutical industry has been completely metric for some time. Prescriptions are filled in centigrams and centilitres instead of in ounces of weight and ounces of liquid. Most scientific laboratories and many engineering laboratories have used the metric system for years. The larger automotive repair garages, and even many small ones, have found it necessary to buy metric tools as well as conventional ones in order to repair foreign-made cars. In the State of Ohio, highway signs now give distances in both miles and kilometres.

Only in the past decade has the drive to go metric gained enough impetus in the United States to push us as close to conversion as we are. The turning point was when England decided to go metric in 1965. That left the United States the only major industrial nation in the world that was not using the metric system, and we are now one of only ten nations of *any* size not using it. The other nine are all small, underdeveloped nations such as Barbados, Gambia, Oman, South Yemen, and Tonga. (When California's Wilson Riles mentioned this on a radio interview, the interviewer asked in mock alarm, "How are we going to trade with Tonga if the United States goes metric?")

The result has been to place American industry at a competitive disadvantage in trading with other nations. Metric nations prefer to buy products made to metric measurements. American industries have been forced to tool some of their plant space to metrics in order to continue to sell to foreign markets, while still maintaining inch-pound machines in order to turn out products for the domestic market. Ford Motor Company, for instance, was turning out 30 percent of its product to metric standards even before it decided on complete conversion.

For this reason alone there can be no turning back. The United States has no choice but to get in step with the rest of the world.

The probability is that within the next ten years the metric system will be the only measurement system taught in our public schools anywhere in the country. It has already been mentioned that California schools will begin teaching the system in 1976. In September of 1973 the Maryland State Board of Education directed the state school superintendent to prepare a plan to convert to the metric system in Maryland schools over a six-year period beginning in the fall of 1974. It can be expected that the other forty-eight states will follow the examples of these two pioneers within a very few years at the most.

Since 1970 numerous bills in favor of metric conversion have been introduced in both houses of Congress. So far only one, HR 11035, has ever been reported out of committee. It was returned for further study, and this time died in committee. Congress as a whole has therefore never had a chance to vote on any of the metric conversion bills introduced in recent years.

The 94th Congress, which convened in January 1975, has several bills under study in committee in both houses.

Since most bills so far introduced have followed a similar pattern, it is probable that the law eventually passed will contain the following clauses:

(1) The change will be voluntary, but the substitution of metric measurement units for customary measurement units will be encouraged in education, trade, commerce, and all other sectors of the economy, with the aim of making metric units the predominant, although not exclusive, language of measurement for transactions within ten years.

(2) Changeover costs shall lie where they fall. (This means that government subsidies will not be paid to industry to compensate for the costs of conversion.)

(3) The federal government will assist in the development of a broad educational program to be carried out in elementary and secondary schools, institutions of higher learning, and among the public at large, designed to teach all Americans to think and work in metric terms.

(4) The term "metric system of measurement" means the International System of Units.

(5) A twenty-five-person National Metric Conversion Board consisting of two senators chosen by the President of the Senate, two congressmen chosen by the Speaker of the House, and twenty-one private citizens named by the President of the United States, will be appointed. Its function will be to develop and submit to the Secretary of Commerce, within one year after the law goes into effect, a comprehensive plan for accomplishing the changeover. In developing the plan, the board will be expected to consult with a broad cross section of American society, including representatives of both large and small industries, businessmen, scientists, engineers, labor leaders, educators, government officials on federal, state, and local levels, and just plain consumers. When appropriate, it will also consult with foreign governments and certain international organizations concerned with metrology (the

science of weights and measures). The plan must be approved by Congress before it can be put into effect.

A treaty signed by the United States nearly a hundred years ago is going to have an effect on our conversion. Because of it we are not merely going to convert to a metric system, but to a specific metric system called the International System of Units.

In 1875 eighteen nations, including the United States, signed the *Convention du Mètre* (the Treaty of the Metre) at Paris. The United States was party to the treaty because, as will be explained in Chapter II, the metric system was legal for use in this country, even though it was not much used. Among other things the treaty established an International Committee on Weights and Measures, to be headquartered in Paris, and under it an International Bureau of Weights and Measures with the responsibility of setting metric standards and of manufacturing and maintaining prototypes (accurate metal models) of the metre and the kilogram.

In 1889 a conference of representatives of the signatory nations was held at Paris to approve the international prototypes the International Bureau of Weights and Measures had called for. Periodically ever since, similar meetings have been held to decide various matters concerning the metric system, There are now forty-two nations signatory to the Treaty of the Metre (including the United States, which is still bound by the treaty). The meetings of the representatives of these forty-two nations have come to have the organizational name of the General Conference on Weights and Measures, usually abbreviated to CGPM (after its French name, *Conférence Générale des Poids et Mesures*). Until 1960 these meetings were held at six-year intervals, but since then have been called by the International Committee on Weights and Measures whenever it decides that a sufficient number of important matters requiring decision have accumulated. There have been three meetings since 1960.

At the eleventh CGPM in 1960 the International System of Units was established. This is customarily abbreviated to SI,

after its French version, *Le Système International d'Unités.*
At the fourteenth meeting of the CGPM, held in 1971, the
most important decision was to add a seventh base unit to
the six previously approved for the system (Base units are
explained in Chapter III.)

The representatives to the eleventh CGPM felt that SI was
necessary because many metric nations, while using the same
basic standards of measurement, sometimes used different
symbols, or even different names for units, and had allowed
their systems to vary from other nations' metric systems in
other ways as well. For instance, some nations used the spellings *metre, litre,* and *gram,* and others used *meter, liter,* and
gramme.

SI lays down rigid rules designed to make the metric system a truly international language of measurement. Most of
these are explained in Chapter V, but in illustration of how
specific these rules are, here are a few:

(1) *Metre, litre,* and *gram* are the only accepted spellings
for those units.

(2) A number never begins with a decimal point. You do
not write .53, but 0.53.

(3) Spaces instead of commas are used in expressing
numbers of more than three digits. You do not write 1,227,430,
but 1 227 430.

(4) Symbols are not to be used as abbreviations. The symbol for metre is not m., but m (no period). Abbreviations have
periods after them; symbols do not. (Many textbooks errone-
ously use symbols as abbreviations.)

If the United States is going metric, it is only good sense to
go all the way and adopt the truly universal language of
weights and measures—SI.

Chapter II

The History of Metrication in the United States

The road leading to metrication in this country has been a rocky one. As a matter of fact, even the road leading to our present system of measurement has been pretty rocky.

Clear back into the dim reaches of history, most nations have legalized the systems of weights and measures in customary use either by royal decree or by legislation. But the Congress of the United States has paid astonishingly little attention to this matter. Although it did once legalize the metre, the litre, and the kilogram, *no law has ever been passed by Congress making the units of the inch-quart-pound system legally designated units.*

In contrast, nearly every other nation in the world has laws designating and precisely defining the units of weight and measure that are legal for use. In some nations, such as France, the use of specified units is compulsory, and the use of units outside the approved system is against the law. This does not mean you would be arrested in France for using a foot rule, but only that items must be manufactured and sold by metric standards and that quantities designated in contracts must be in metric units.

The seeds of the metrication controversy in the United States were sown more than 150 years ago, in 1821, by then Secretary of State John Quincy Adams, but even before that a running debate on the general subject of measurement systems was going on.

Under the Articles of Confederation of 1777 the Continental Congress was given the authority of "fixing the standard of weights and measures throughout the United States." It never got around to passing any legislation on the subject, though, so the traditional English units that the colonists brought with them when they came to America remained in general use without actually having any legal sanction.

The Constitution of the United States, which superseded the Articles of Confederation in 1789, similarly authorized Congress to "fix the standard of weights and measures." In his very first message to Congress, on January 8, 1790, President George Washington urged legislative attention to the matter.

As a result of Washington's suggestion, the House of Representatives asked Thomas Jefferson, then Secretary of State, to prepare a plan for consideration. Jefferson responded with two alternative plans. One was merely a plan to "define and render uniform and stable" the weights and measures already in use; the other was a new system of his own invention based on decimals. This was not the metric system, although by coincidence the Paris Academy of Sciences began to create the metric system, which also had a decimal base, that same year. In the Jefferson system inches and pounds were still the basic units of measurement, but there were 10 inches in a foot and 10 ounces in a pound.

Congress debated on both plans for years, but in the end failed to adopt either.

It was not until 1799 that Congress got around to passing any legislation at all concerning weights and measures. The Surveyor Act of that year required the surveyor of each port to test the weights, measures, and instruments used in collecting custom duties at least twice a year. Unfortunately, since Congress had set no standard, the act could not be put into effect.

In 1807 Ferdinand Hassler, upon being appointed to head the new Survey of the Coast (later to become the United

States Coast and Geodetic Survey) decided to use the metre as the standard length for that work. Although this was merely an administrative decision, without the legal force of a legislative act, its consequence was that the first official measure ever used by the United States government was the metre.

In 1817 an attempt was made in Congress to revive discussion of Jefferson's two plans, but since these were now twenty-six years old, too many members of Congress felt that an entirely fresh approach was needed. In the end the Senate passed a resolution asking the Secretary of State to prepare a new statement "relative to the regulations and standards for weights and measures in the several states."

Secretary of State John Quincy Adams responded with a study that was so exhaustive it took four years to prepare. It was presented to Congress in 1821.

By then the metric system devised in France was well known throughout the world. Except in a few small European countries, however, it had never been in general use outside of France, and at this particular time it was not even in exclusive use there anymore. In an 1812 decree that remained in effect until 1837, Napoleon Bonaparte had suspended the compulsory provisions of the 1795 metric-system law and had restored the prerevolutionary unit names and values for French weights and measures. Temporarily both systems were in use in France, therefore. Nevertheless, Adams was so impressed by the system that he included a detailed study of it in his report. The report thoroughly discussed the advantages and disadvantages of both the English system and the metric system for the United States. Despite Adams' conclusion that the metric system "approached the ideal perfection of uniformity applied to weights and measures," he counseled against adopting it at that time because most of the country's trade was with inch-quart-pound England. Adams felt that eventually it might be wise to adopt metrics, or some other uniform international measurement system acceptable to the

nations with which the United States traded, but that at the
moment it would be unwise to adopt a system different from
that of our main customer.

Similar thinking has fueled the controversy over going metric ever since. While it is true that some opponents have based
their objections on arguments against the metric system itself,
far more have simply objected to the idea of *conversion*. Like
John Quincy Adams, many recognized the merits of the system, but felt that the cost and confusion of a changeover
would be too high a price to pay, even if the metric system
was more efficient.

A much-cited quotation from General John J. Pershing illustrates this type of thinking. At the height of the controversy
both the pro- and antimetric forces were not above bending
the truth in order to score points. In the period immediately
after World War I, a favorite device of the pro forces was to
quote the following passage from a letter written by General
Pershing in order to show that the Commander of the American Expeditionary Forces of World War I was pro-metric:

> The experience of the American Expeditionary Forces in France
> showed that Americans were able readily to change from our exist-
> ing weights and measures to the metric system. I think the principal
> advantages of the metric system are summed up in the fact that
> this is the only system which has a purely scientific basis. Not the
> least advantage of the fact that the metric system is based on
> scientific principles is the facility which that system gives to cal-
> culations of all kinds, from the simplest to the most complex.
>
> I believe that it would be very desirable to extend the use of the
> metric system in the United States to the greatest possible extent.

This passage was accurately quoted, but there was more to
the letter. What the pro-metric advocates never mentioned
was the next paragraph:

> But I can readily see that there would be many practical ob-
> stacles in the attempt entirely to replace our existing system by

the metric. These obstacles have to do especially with manufacturing plants and with existing records of all kinds. I am not sufficiently familiar with the technical phases of the question to be able to say whether or not such obstacles might be overcome, and as a consequence I would prefer not to be quoted as advocating the replacing of our present system by the metric system.

John Quincy Adams' exhaustive report brought no more action from Congress than Thomas Jefferson's previous report had. But in 1828 Congress did adopt the troy pound (a pound of 12 ounces customarily used for weighing gold and silver) as the standard for the United States mint in coining money. Thus after fifty-one years of existence, the United States finally had at least one official weight standard, but it applied only to coinage.

Two years earlier, in 1826, a flurry of debate had been stirred in Congress when a House investigative committee disclosed that variations in custom house standards in different parts of the country were costing the United States Treasury considerable loss of revenue. Because there was no official standard, each custom house set its own, and since there happened to be several different measurement systems in common use, it was quite natural that every custom house did not choose the same one. There was, for instance, the United States foot, the Bushwick foot, and the Williamsburg foot, all of different lengths. There was the avoirdupois pound of 16 ounces and the troy pound of 12. There was the imperial bushel and the Winchester bushel. Furthermore, because the weights and measures in use had not been manufactured from a single prototype, or model, even those purporting to be of the same system often varied from each other.

As usual the debate resulted in no legislation, but when the subject was revived four years later, in 1830, the Senate passed another resolution. This one directed the Secretary of the Treasury "to cause a comparison to be made of the standards of weights and measures now used at the principal custom houses in the United States and report to the Senate at the next session of Congress."

In 1818 Congress had abolished the Coast Survey and had turned the survey work over to the Army and Navy, putting Ferdinand Hassler out of a job. In 1832 they reestablished it again on the basis of the original act of 1807, and Hassler again became its superintendent. In the interim he held a number of other jobs, and very nearly impoverished himself in a disastrous business venture. In 1830 he was working as a gauger in the New York Custom House.

Secretary of the Treasury Samuel Ingham appointed Hassler to make the comparison ordered by the Senate. Hassler's report, showing large variations in the weights and measures in different custom houses, was presented to Congress in 1832.

Meantime Louis McLane had replaced Ingham as Secretary of the Treasury. He interpreted the Senate resolution of 1830 as giving him authority to straighten out the mess in the custom houses without waiting for legislation. By administrative decree he declared that custom house standards of measurement from there on would be the yard, the avoirdupois pound, and the Winchester bushel (a bushel of 32 dry quarts, which later became the United States standard bushel). He also directed Hassler to call in all the weights and measures in use in the custom houses and to issue new ones that were all alike.

We now had official standards at least in United States custom houses, although their legality might have been in doubt if anyone had questioned them, because they had been set by administrative decision instead of by law. Nothing more happened until 1836, when a joint resolution of the Senate and House directed the Secretary of the Treasury

to cause a complete set of all weights and measures adopted as standards and now either made or in progress of manufacture for the use of the several custom houses, and for other purposes, to be delivered to the Governor of each State in the Union, or such person as he may appoint, for the use of the States, respectively, to the end that a uniform standard of weights and measures may be established throughout the United States.

Another resolution directing that the states be furnished with weighing balances was passed in 1838. So now nation-wide standards were in effect, not by legislation, but merely by administrative order from the Secretary of the Treasury. And his authorization to issue such orders was not even a statute, but merely a couple of Congressional resolutions.

With the 1838 resolution, Congress seemed to decide it had met its Constitutional responsibility to "fix the standard of weights and measures." It made no more efforts in that direction until 1866. Then it quietly dropped a legislative bomb-shell by passing House of Representatives Bill 596. The bill, which passed with little debate and with even less public notice, declared that it was lawful throughout the United States "to employ the weights and measures of the metric system." It specified that "no contract or dealing, or pleading in any court" could be deemed invalid because of use of metric measurements. The bill was approved by the Senate and was signed into law by President Andrew Johnson on July 28, 1866.

The new statute merely *permitted* the use of the metric system in the United States; it did not adopt the system as the official standard. But since no other standard had ever been legislated by Congress, the only legally recognized system of weights and measures in this country was the metric system. As a matter of fact, it still is.

Some background information is necessary in order to understand how H.R. 596 came to be passed.

In 1837 a French law reinstated the metric system and made it compulsory throughout France (although the law did not go into effect until January 1, 1840). From 1840 on, the metric system began to receive ever-widening acceptance, particularly in the scientific community, which recognized in it a much-needed universal language of science.

In 1843 a well-known scientist and educator named Alexander Bache, a great-grandson of Benjamin Franklin, was appointed to succeed Ferdinand Hassler as superintendent of the Coast Survey. By then an Office of Weights and Measures had been placed under that bureau, which made Bache responsible for the supervision of national measurement standards.

Alexander Bache had studied in Europe from 1836 to 1838 and had become sold on the metric system. He became its tireless advocate, and gradually he converted everyone with whom he had contact to his point of view: the thirteen different Secretaries of the Treasury under whom he served during his long career, the Senate and House leaders, and everyone else who would listen.

One of his most important conversions was Congressman John Kasson of Iowa, Chairman of the House Committee on a Uniform System of Coinage, Weights and Measures, which was created in 1864. Kasson became nearly as tireless a pro-metric exponent as Bache, and was in a strategic position to do something about it. It was Kasson who pushed the bill through the House.

It probably would be an exaggeration to give Alexander Bache sole credit for the metric system becoming legal in America, but without his constant propagandizing, the act might never have come about. He died a year after its passage, still in office, at the age of sixty-one.

While the seeds of controversy over the metric system were sowed by John Quincy Adams in 1821, they really did not begin to flower into a public controversy until after the passage of H.R. 596. Prior to that, what debate there was over the metrication issue had been pretty well confined to the halls of Congress, but now it became a public issue.

One reason was that the statute really did not resolve anything. Since the law did not make the metric system the only permissible system, as the 1837 law in France had done for that country, legalizing metrics merely added one more system to the confusing number already in use. In actual practice, few people outside of scientific laboratories used it.

Now that the metric system was legal, the next goal of the pro-metric advocates was to educate the public to use it. The keynote was set by Congressman Kasson in an address to a New York teachers' group less than a month after the statute became law. He urged the teachers to make a strenuous effort to teach the metric system to "the rising generation."

Kasson's plea generated considerable interest from other metric-system advocates, and before long, extensive propaganda campaigns to induce the public to use the metric system were being launched by various groups. Because of the character of human nature, advocacy of *any* cause inevitably brings opposition from some source. Former New York State Governor Al Smith once said that he could easily form a group of New Yorkers favoring the resale of Manhattan back to the Indians for twenty-four dollars, if only someone would first organize a group advocating that it not be sold back. As might have been expected, the pro-metric campaign quickly resulted in the formation of a number of antimetric groups. Both sides began to send speakers on the lecture-tour circuit and to issue streams of pamphlets pressing their respective views.

Most of these organizations were merely local groups and were short-lived. The lines of battle were essentially drawn by three national organizations reflecting the differing personalities of two men.

The American Metrological Society, formed in 1873, was an association of scientists and educators concerned with the science of weights and measures in general, and not just with the metric system. It supported the general adoption of that system, however, and issued a number of scholarly pamphlets on the subject.

The American Metric Bureau was formed in 1876 and had the sole purpose of getting the metric system adopted as the

standard for the United States. Although these were two separate organizations, they had a common president in Frederick
Barnard, who was also president of Columbia University.

The advocacy campaign of both organizations was geared
toward public education and was generally restrained and
reasoned. Neither was ever guilty of such truth-bending devices as the excerpted quotation from General John Pershing's
letter.

The first organized antimetric society of national scope in
the United States was the International Institute for Preserving and Perfecting Weights and Measures, founded in 1879.
This organization was headed by a Cleveland, Ohio, engineer
named Charles Latimer. Its avowed purposes were to lobby
against any legislation proposed to further the metric system,
to preserve the use of the inch-quart-pound system, and to
promote the rather fuzzy mystic concept that the dimensions
and construction of the ancient pyramid-tomb of King Khufu
at Giza, one of the Seven Wonders of the World, proved the
heavenly origin of certain scientific laws, including standards
of measurement.

In 1880 Latimer published a book titled *The French Metric
System, or, The Battle of the Standards*. The book's preface
gave its purpose as "the awakening of the advocates of the
French system to the defeat that lies before them." These
paragraphs from near the beginning of the book are indicative
of its tone:

The followers of Darwin, and the infidel will both deny the in-
spiration of our weights and measures, and ascribe all of our
progress to a natural progression; and, doubtless, will hail the appearance of the new French unit as another argument in favor of
their peculiar views and theories, and will be equally ready to
re-adopt the fantastic freaks of the French Revolution, even to
abandoning the Sabbath and burning the Bible. . . .

It may be thought by some unreasoning persons that there has
been so much said and done with reference to the French metric
system, that there now remains nothing more to be said or done

but for Congress to issue its edict, and that thereupon the French metric system will be at once an accomplished fact and the law of the land. . . . To these gentlemen it may be well to say . . . How dare you attempt to foist upon us without our consent new weights and measures unknown to us and to our fathers? Understand that we will, with one blast of our mouth, cast down your false measure.

Charles Latimer died in 1888, and a year later the prime pro-metric exponent of the day, Frederick Barnard, also died. Without their leadership both the International Institute for Preserving and Perfecting Weights and Measures and the American Metrological Society became virtually inoperative. (The American Metric Bureau had quietly expired some years earlier.) With the deaths of the spiritual leaders of the two factions, the controversy temporarily came to an end. No champions for either side arose immediately to take their places, and the matter faded as a public issue until after the turn of the century.

Before the controversy resumed, a number of events occurred which affected the history of weights and measures in the United States.

The 1875 Treaty of the Metre has already been mentioned, as well as the first General Conference on Weights and Measures held in 1889 to approve the prototypes of the metre and kilogram built under the supervision of the International Bureau of Weights and Measures. In 1890 the United States' copies of the prototypes were delivered to the White House, where they were accepted in a formal ceremony by President Benjamin Harrison.

In 1893 United States Superintendent of Weights and Measures Thomas Mendenhall announced that these prototypes were not only the United States' standards for the metric system, but henceforth would be considered the nation's "fundamental standards of length and mass." This meant that thereafter units of the customary system would not be defined by their own standards, but in terms of the metric units. The yard would be defined by specifying what fraction of a metre made a yard, and the pound would be defined by specifying what fraction of a kilogram made a pound. (The United States Bureau of Standards' official definition of a yard is still 3600/3937 of a metre.)

In 1894 a law establishing metric units for electric measurement was passed by Congress. In 1901 Congress established the National Bureau of Standards.

The metrication controversy resumed in 1902, when the House Committee on Coinage, Weights and Measures held a series of hearings on a proposal to adopt the metric system as the only legal standard for the United States. The first witnesses, mainly scientists and industrialists, were mostly in favor of the adoption, but then opposition developed. A number of small manufacturers appeared before the committee to protest that the cost of conversion would work a hardship on them. *The New York Times* published a sharp editorial deploring "the pernicious activity of those who are trying to crowd the metric system of weights and measures upon the country." And then Frederick Hasley appeared.

Hasley and Samuel Dale's book, *The Metric Fallacy,* has already been quoted at the beginning of Chapter I. It had not yet appeared in 1902, but largely through the opposition stirred up by Frederick Hasley, the proposal to adopt the metric system was tabled by Congress. When it arose again in 1904, and another series of public hearings lasting until 1906 was held, the publication of Hasley and Dale's book in 1904 was a factor in again getting the proposed legislation tabled.

Because the pro-metric forces temporarily gave up after that, the controversy again died down until 1916. In that year the American Metric Association (still in existence today, but now called the Metric Association) was formed. Its goal was to obtain general acceptance of the metric system, and its strategy was to attempt to gain the support of scientific, educational, and professional organizations.

That same year Frederick Hasley and Samuel Dale organized the American Institute of Weights and Measures to fight the American Metric Association.

America's entry into World War I suspended the metric battle until 1919. In that year the World Trade Club was formed in San Francisco. Financed by a wealthy manufacturer named Albert Herbert, the organization maintained a full-time lobbyist in Washington, D.C., to push for legislation to adopt the metric system. It also issued a constant stream of pro-metric propaganda. In 1920 this group changed its name to the World Metric Standardization Council, and in 1924 again changed it to the All-American Standards Council, but its membership and tactics remained essentially the same under all three names.

In 1920 a campaign to flood Congress with pro-metric post-cards was launched by the World Trade Club, resulting in more than a hundred thousand cards being mailed to congressmen. In reaction, a countercampaign was initiated by Nathan Viall, editor-in-chief of *American Machinist* magazine, of which Frederick Hasley was an associate editor. Simultaneously, Hasley published a revised edition of *The Metric Fallacy* (over the objection of his collaborator, Samuel Dale, who was still as antimetric as ever, but thought the revision would not sell; Dale's name as coauthor did not appear on this edition).

The activities of Viall and Hasley effectively counteracted those of the World Trade Club under its various names, and as usual, no legislation was passed.

From 1926 to 1931 agitation for adoption of the metric system progressively declined. In 1931 the All-American Standards Council (the final name of the World Trade Club) folded up. By then the American Institute of Weights and Measures had virtually disappeared also. It disappeared completely in 1935, when both Frederick Hasley and Samuel Dale died.

With the demise of the American Institute of Weights and

Measures, effective organized opposition to metrication never again developed. For twenty-six years there was really nothing to oppose, because there was also no organized effort to get metric legislation passed. The only important pro-metric society still in existence was the Metric Association, and it mounted no major campaigns after 1931.

In 1957 the United States Army issued a regulation requiring metric standards for all its weapons and equipment. The regulation barely received mention in the news.

In 1968 Congress passed the Metric Study Act, directing the Secretary of Commerce to arrange for a three-year study to determine the impact of the steadily increasing use of the metric system on America. The law required the Secretary, on the basis of the findings, to make "such recommendations as he considers to be appropriate and in the best interests of the United States."

In July 1971 then Secretary of Commerce Maurice Stans released a 170-page report to Congress, the conclusion of which is indicated by its title: *A Metric America: A Decision Whose Time Has Come.* His recommendation, based on the findings, was that "the United States change to the International Metric System deliberately and carefully; that this be done through a coordinated national program; and that early priority be given to educating every American schoolchild and the public at large to think in metric terms."

Twelve supplemental investigations have since been made on specific aspects of metrication, including such matters as its probable impact on industry, education, and foreign trade, and its effects on consumers. It is on the basis of these studies that the metrication bills now before Congress were drafted.

So far no organized opposition of any appreciable size has developed.

Chapter III

How Our Customary System Developed

When engineers refer to the system of measurement in general use in the United States today, they usually call it the inch system. In trade and commerce it is called the inch-pound system or, sometimes, the inch-quart-pound system. (Actually, of these three units, only the pound is a base unit. Base units are explained a few paragraphs farther on. The yard and gallon are the generally recognized base units of length and liquid capacity.) Educators variously refer to the system as the English system, the English–United States system, and the customary system.

For purposes of simplification, it will be referred to hereafter as the customary system.

The story of how the system developed literally goes back to the Stone Age, to the first caveman who realized that a club just *so* long was superior to one either longer or shorter. But before we get to that story, it is necessary to understand a few things about measurement in general.

All systems of measurement, no matter how much they differ from each other in detail, have four basic concepts in common:

(1) The things to be measured are called *quantities*. Quantities are abstractions such as length, volume, weight or mass, time, and temperature.

(2) For each quantity there must be a *base unit*, which is simply the basic value by which the quantity is measured.

The base unit of the quantity *length* in the customary system is the *yard*. The base unit of the quantity *volume* is the *gallon* when referring to liquid volume and the *bushel* when referring to dry volume. The base unit of the quantity *weight* or *mass* is the *pound*. The base unit of *time* is the *second;* that of *temperature,* the *degree Fahrenheit.*

(3) Base units may be divided or multiplied to form other units, smaller or larger. Subdivisions of the yard are the *foot* and the *inch;* multiples are the *rod* and the *mile*. Subdivisions of the gallon are the *quart* and the *pint;* multiples are the *barrel* and the *hogshead*. Subdivisions of the bushel are the *peck* and the *dry quart;* its multiple is the *dry barrel*. Subdivisions of the pound are the *ounce,* the *dram,* and the *grain;* multiples are the *hundredweight* and the *ton*. The only subdivisions of the second are *fractions of a second,* but its multiples are the *minute,* the *hour,* the *day,* and the *year*. Subdivisions of the degree Fahrenheit are expressed in decimal fractions; multiples are merely expressed numerically, as, for example, 20 degrees.

(4) In modern times each unit has a *standard*. The purpose of standards is to make sure each specific measuring unit in use is exactly the same size—that is, that all yardsticks are the same length, that all pound weights are the same mass, that all gallon containers hold precisely the same amount of liquid. In actual practice, of course, most of the measuring devices in common use are slightly off one way or the other, because extreme accuracy of measurement is not required for most everyday purposes. But very accurate measurements are often required in the manufacturing industries, and extremely accurate measurements are usually required in engineering and scientific laboratories. Therefore standards are maintained so that the units in use can be checked against the standards for accuracy. A standard may be merely a definition. The standard for the nautical mile is 1 minute, or 1/60 degree of the earth's circumference at the equator. More often a standard is an actual object maintained under a precise set of circum-

stances. The standard for the yard in this country, for instance, is a metal bar at the National Bureau of Standards that is always kept at the same temperature, because metal expands and contracts with changes in temperature.

Unfortunately, in metrology the word *standard* is also used as a synonym for *system*. In the Constitutional authorization to Congress to "fix the standard of weights and measures," the word is used in this sense. Actually, the United States does not have a standard of weights and measures, because Congress has never fixed one, but in France the standard is the metric system. If the distinction between these two separate uses of the word *standard* is kept clearly in mind, there should be no confusion.

To get back to the story of how our customary system developed, its origins lie in the dawn of civilization. In recent years archaeologists have come to the conclusion that civilization began in the lower part of the Tigris–Euphrates valley of Southwest Asia. The Sumerian culture of that area is judged to go back at least 6,500 years. The Sumerians invented the first numerical system, a complex system that used alternate multiples of 10 and 6, to produce such progressions as 10, 60, 600, 3600. Not much is known about the Sumerians' earliest system of measuring length, volume, and weight, although a Sumerian cubit ruler about 10½ inches long dating about 2000 B.C. was found by archaeologists some seventy-five years ago, but one of their other measurement units is still in use today. It was the Sumerians who divided the circle into 360 degrees.

Scholars generally agree that the first quantity measured by prehistoric people must have been length, and that the first linear units were based on parts of the human body. The thickness of a finger is known to be the origin of the digit, a unit of ¾ inch that probably came from the Sumerians, but whose first known use was by the ancient Egyptians. The digit is still used as a unit of measurement in some rural areas of Great Britain. The width of a man's thumb is the origin of the inch, and, of course, the length of a man's foot is the origin of the foot. In ancient Egypt the cubit (also a unit of linear measure used by other early civilizations and mentioned several times in the Bible) was the distance from the tip of the middle finger, when the hand was outstretched, to the elbow. The hand (4 inches) is still the unit used for measuring the height of horses.

Volume was probably the second quantity that primitive man found necessary to measure, and the logical probability is that the first measuring devices used for volume were receptacles found conveniently at hand, such as seashells, gourds, and the skulls of large animals.

The measuring of weight is a much more sophisticated process. It is probable that the first crude attempts at weighing consisted merely of hefting the item to be measured in one hand and hefting stones of various sizes in the other, until the item and the stones felt about the same weight. We do not know when some primitive experimenter first thought of tying a cord about the middle of a stick and suspending a container from either end, but we do know that the balance scale predates even the Sumerians. A prehistoric Egyptian balance with limestone weights, which is estimated to be seven thousand years old, was unearthed by archaeologists not long ago.

Archaeological evidence indicates that weighing among ancient peoples was largely restricted to precious metals and gems. For most commercial dealings, either capacity measures were used, or items were simply counted. The latter is still a common method of selling commodities. In recent years grocery stores have tended to sell fewer and fewer things by the dozen—oranges, apples, and bananas are now usually sold by the pound, for example—but there are still items such as eggs and doughnuts that are sold no other way.

Early balance scales were probably used primarily for weighing precious metals and jewels because these were the common mediums of exchange before money was invented. Barter was the method of trade, and it was much easier and more convenient for a man who was traveling to a neighboring village in order to purchase an ox to carry along a small bag of gold, silver, or copper—in ancient times copper was a semiprecious metal—than to drive half a dozen goats ahead of him.

About 700 B.C. the practice of stamping the weights of bits of gold, silver, and copper on them in order to make such transactions easier developed in the kingdom of Lydia in Asia Minor. This is how coinage was invented. Because the Mesopotamian weight unit of the *mina*, its subdivision the *shekel*, and its multiple the *talent*, were in common use throughout the Mediterranean area, the first coins took the names of those weights.

Ever since, throughout history, there has been a tendency to associate weight and money units. The English *pound* is the most obvious example. The basic monetary unit in France before the franc replaced it was the *livre*, which means *pound*. The *lira*, still the basic monetary unit of Italy, is derived from *libra*, the Latin word for pound.

Before coinage was invented, great care was taken to make sure weights were accurate, because precious metals were, in effect, money. Merchants carried their own scales and weights with them, and it was common practice to check the other fellow's measure on one's own scale before closing a deal.

As a result of this practice, weights in ancient Egypt and Mesopotamia were remarkably accurate. A recent examination at Yale of sixteen Egyptian 1-kedet weights from about 1000 B.C. showed that the lightest weighed approximately .31 ounce, the heaviest .34 ounce, and that most were within 1/100 ounce of each other.

In ancient times this kind of standardization was not as common for units of measure other than weight, though. Because the width of people's fingers varied, the length of the digit varied from place to place. Obviously, thumbs were not all the same width either; nor were all men's feet the same length or the distances from their center fingertips to their elbows all equal.

Once civilization had advanced enough for men to build boxes and basket containers for dry goods, and to make crocks and jugs for liquids, capacity measures no longer had to be the odd sizes of whatever was available, such as a seashell or an animal skull, but could be built to any size desired. The tendency was to relate their sizes to linear units, creating such volumes as the cubic foot and the cubic cubit, but since linear units varied from place to place, so did volume units.

The variation of measurement units from locality to locality, sometimes even in the same country, has been a problem that metrologists have wrestled with throughout history. An attempt was made to standardize units as early as 2800 B.C. The Egyptian Pharaoh Khufu, builder of the Great Pyramid at Giza, is believed to have set a standard for the Egyptian cubit. The belief is not based on actual record, but upon the assumption that there must have been an accurate standard in order to account for the amazing accuracy of measurement in the building of the pyramid. The average linear difference between the four 755-foot sides is only 1/4000 of their length.

Hammurabi, king of Babylon about 1700 B.C., is believed to have set standards for both length and weight. There is no historical record of standards in ancient Greece, but modern measurements of more than a dozen classical Greek buildings indicate that a foot measure of 12.45 inches was used in the construction of all of them, with an average variation of only 6/100 of a foot. Such accuracy would have been impossible without a standard foot.

If we judge by the Old Testament, there were standards in ancient Israel for weight and volume, but it was not unknown for merchants to carry two separate sets of weights and measures, one that weighed and measured larger than the standard, to use for buying, and one that weighed and measured smaller, to use when selling. Deuteronomy 25:13–15 reads:

Thou shalt not have in thy bag divers weights, a great and a small. Thou shalt not have in thine house divers measures, a great and a small. But thou shalt have a perfect and just weight, a perfect and just measure shalt thou have: that thy days may be lengthened in the land which the LORD thy God giveth thee.

Proverbs 20:10 reads:

Divers weights, and divers measures, both of them are alike abomination to the LORD.

The Greeks adopted, with little change in value, but with changes in nomenclature, the Mesopotamian units of measurement that had spread throughout the Mediterranean area. The Greek *drachma*, both a coin and a unit of weight, was fixed by law about 600 B.C. at 1/100 of the Mesopotamian mina, for example. (Later the Roman pound was ½ mina.)

Greek measurements passed to Rome. Again there were changes in the names of units, but little change in their values. Rome made few innovations in the science of metrology, but it did contribute to the development of our customary system in three ways.

First, through military conquest and commercial trade, the Roman Empire spread a fairly uniform system of measurement throughout the civilized world. Second, the Romans adopted a duodecimal system (that is, a system based on divisions and multiples of twelve) that is still the basis for some parts of the customary system, such as the 12 ounces in a troy pound and the 12 inches in a foot. Third, a good deal of the nomenclature in the customary system derives from Latin. The modern abbreviation for pound, lb., comes from *libra*, meaning "pound." Both the *libra* and the *pes* (the Roman foot) were divided into *unciae*, which simply means twelfth parts. From *unciae* come both our *inch* and *ounce*. The word *mile* comes from *mille passus*, which meant 1,000 paces.

The Roman pace was 5 Roman feet, or the distance a soldier moved in two strides. The *mille passus* was therefore

5,000 feet, which makes it close to the modern statute mile of
5,280 feet, since the Roman foot was less than .4 inch shorter
than our customary foot.

Historians usually date the fall of the Roman Empire as A.D.
476, because that was when the last Roman emperor, Romulus
Augustulus, was overthrown by the barbarian Odoacer, who
assumed the title of King of Italy. However, this was merely
the final incident in a gradual decline of power that had been
going on for more than 280 years.

The so-called barbarian invasion of the Roman Empire was
no sudden onslaught. It might be said to have begun as early
as A.D. 193, when Rome started importing large numbers of
German mercenaries to fill the gaps in its military ranks because it could no longer induce a sufficient number of Roman
youths to enter military service.

Other Germans, who heard about the wonders of Roman
civilization from returning mercenaries, migrated there. The
first migrations were entirely peaceful, taking place with the
permission, and sometimes even the encouragement, of Rome.
One of the largest migrations occurred toward the end of the
fourth century, when the Goths, who were driven westward
by the Huns, a nomadic Mongolian people, were given blan-
ket permission to settle in the Balkans. By the time the Visigoths, the Vandals, and the Huns began their sweeping raids
in the fifth century, Germanic people were already settled
throughout the empire. As the invaders conquered areas, some
of them settled down and merged into the conquered com-
munities, intermarrying with the Romans.

The Germanic people who had migrated to Rome brought
many of their own customs with them, including their own
system of measurement. The invading barbarians brought not
only their own customs, but their own law, which in some
places usurped Roman law and in others mingled with it to
form entirely new legal systems. Since throughout history
measurement units have customarily been fixed by law (except in the United States), these changes in the legal system

often changed measurements. A few examples will give some idea of the variety of the sources of these changes: The Franks and Burgundians took their tribal laws from western Germany to France; the Visigoths took theirs to Spain; the Lombards settled in Italy; the Anglo-Saxons overran England.

Then, in the tenth century, another source of change was the great empire of Islam, which spread Arabic numerals across Europe. It was nearly two hundred years before this numerical system entirely replaced the Roman system; and in many localities Roman numbers were still in use while villages only a few miles away had adopted the use of Arabic numerals.

When the feudal system of the Middle Ages began to develop, the impact of all these divergent forces on the measurement systems in Europe was catastrophic. Each feudal baron became responsible for the law in his own territory, and thus had the power to decree what the measurement standards would be. Voltaire once said that when a man traveled across France, "the law changed as often as he changed horses." Measurement units changed as often as the law.

The names of units tended to remain the same within each language, but standards varied so much that the foot in Europe ranged all the way from 10 inches to 20 inches. A thirteenth-century French contract concerning the sale of some land has been discovered in which the parties found it necessary to specify the length of the foot by which the land would be measured. This was probably common practice. The number of ounces in a pound ranged from 6 to 18, depending on where a person was.

Sometimes several different weight systems were in use in the same locality. This stemmed from the medieval belief that different materials should be weighed by different standards. As a rule precious metals and gems were weighed by pounds with fewer ounces, possibly for the psychological effect of making the buyer think he was getting more than he actually was. This philosophy has extended to modern times with the general use of the troy pound of 12 ounces to measure gold and also with the use of the short and the long ton. For many years American coal miners were paid on the basis of the number of long tons of 2,240 pounds they dug, then the coal was sold in short tons of 2,000 pounds. Another survival from the Middle Ages is the carat, the unit of weight for diamonds and the unit of fineness for gold. Initially, the carat was the weight of a single carob seed.

The most chaotic area of medieval measurement was in units of capacity. There were such units as the *sack,* the *bundle,* the *firkin,* the *stake,* the *pottle,* and the *cartload,* all varying in amount from place to place. To add to the confusion, some units of volume also became units of weight. Our present-day ton provides an example of how this sort of thing came about. Originally, *tun* or *tunne* was an Anglo-Saxon word for tub or vat. From that origin the ton became a wine measure of 252 gallons. Since that amount weighed approximately 2,000 pounds, the ton also became a measure of weight.

Coinage, weight, liquid measure, and dry measure were all confusingly interrelated in an edict called the Assize of Bread and Ale, issued by King Henry III in 1266. In part it read:

An English penny called a sterling, round and without any clipping, shall weigh thirty-two wheatcorns in the midst of the ear; and twenty pence do make an ounce, and twelve ounces a pound; and eight pounds do make a gallon of wine, and eight gallons of wine do make a bushel, which is the eighth part of a quarter.

By *quarter* Henry meant a quarter of a ton. So if you reread the above passage, you will realize that it starts with coinage, defines the coinage in terms of weight, goes back to coinage, again defined in terms of weight, then defines that weight in terms of liquid measure, which in turn is defined in terms of dry measure, and then goes back to weight for the final definition of the dry measure.

Even though the Assize of Bread and Ale is confusing, it

represented a trend of the period for rulers to attempt to bring some sort of order to metrology. Near the end of the eighth century, Charlemagne had attempted to set uniform standards of weights and measures throughout his empire, but local custom had been too firmly rooted. By the thirteenth century, though, as the demise of the feudal system began and rulers began to enjoy more complete control over their realms than had been possible under the loose federations of feudalism, most European rulers began not only to set weights and measures standards, but to enforce them. This did not take place everywhere in Europe, because it took another two hundred years for feudalism to die out completely, but it was the general trend.

Standards were seldom chosen arbitrarily; usually they were based on some factor that seemed logical to the standard setter at the time, although they often seem illogical from the more scientific point of view of today. Henry III no doubt thought that thirty-two wheatcorns taken from the middle of an ear was quite an accurate standard. A hundred and fifty years before that, Henry I decreed that the English yard was the distance from the tip of his nose to the end of his thumb when he held his arm straight out from the shoulder. Before that the *girth*, or yard of the Anglo-Saxons, was the length of the king's waist sash.

Henry III's successor, Edward I, was responsible for another stupidity in our customary system: the long ton. The British hundredweight, as might be suspected from its name, was 100 pounds. Edward issued a decree changing the hundredweight to 112 pounds. In case that was not confusing enough, he ordered that a temporary hundredweight of 108 pounds be used until a specified date, when the 112-pound hundredweight would go into effect. The latter is the basis of the present-day long ton of 2,240 pounds.

A few more standards set in England at various times were:

(1) The inch: the width of three barleycorns, round and dry, laid together.

(2) The foot: decreed in 1305 by Edward I to equal 12 three-barleycorn inches.

(3) The rod: originally a pole of 15 Saxon feet used to measure real estate; when Edward I changed the standard of the foot from the Saxon length of 13.2 inches to 12 inches, the rod remained the same length, but became 16.5 feet.

(4) The furlong: the customary length of a furrow plowed by an ox-drawn plow, or 40 rods.

(5) The acre: the amount of land a farmer could plow in a day, or 40 rods by 4 rods.

(6) The grain: the weight of a grain of wheat taken from the middle of the wheat ear.

Later, standards became a little more accurate, but not much. In 1490 Henry VII had an octagonal iron bar made as the standard of the yard; but it was crudely made, and its foot and inch subdivisions were not even equal. A hundred years later Queen Elizabeth I replaced it with a brass rod a half-inch in height and depth and a yard long. According to metrologist Francis Baily, who examined this rod in the British archives in 1836, its accuracy was not much better. He wrote: "A common kitchen poker, filed at the ends in the rudest manner by the most bungling workman, would make as good a standard."

In 1758 a committee of Parliament was appointed to set standards for weight and measure. Under its direction a new yard bar and a troy pound prototype were manufactured with the most accurate tools and instruments of the day. These prototypes were deposited with the clerk in the House of Commons, and the committee report urged Parliament to adopt them as official standards.

Seventy years later, in 1828, Parliament finally got around to adopting the suggestion. The "new" prototypes were declared the only "original and genuine" standards for the yard and pound.

It had not occurred to anyone to provide for their safekeeping, though. They were still lying in the office of the clerk of the House of Commons when the Houses of Parliament were destroyed by fire in 1834. The prototypes burned up along with everything else.

In 1838 another parliamentary committee was appointed to create new standards. There had been considerable scientific advancement in the eighty years since the destroyed standards were built, and the new standards were constructed to such great accuracy that they remained the official standards of length and mass for Great Britain until the metric system was adopted in 1965. The mass standard, which was made of platinum, was not the troy pound this time, but the avoirdupois pound of 16 ounces.

The Weights and Measures Act of 1878 officially established these standards. A number of copies were made, and two of the yard bars were sent to the United States, where they were used as our national standard of length until 1893, when United States Superintendent of Weights and Measures Thomas Mendenhall decreed that the metre-bar prototype sent to us from France would thereafter be the fundamental standard of length.

The history of the development of the customary system used in the United States is essentially the history of its development in England. With minor exceptions, such as the difference between the English imperial gallon and the United States gallon, it was transferred virtually intact from the mother country.

Chapter IV

Metric-System Advantages

Unlike other measuring systems, the metric system did not gradually develop from custom and usage. It was invented whole as a new system based on the latest scientific principles of the day, and was deliberately designed to replace the customary system in use in France at the time.

Historians differ as to who deserves the credit for first conceiving of the system. Flemish mathematician Simon Stevin suggested the adoption of measurement units with a decimal base as early as 1585. A more popular choice among historians is Gabriel Mouton, the vicar of St. Paul's Church in Lyons, France. In 1670 he devised a decimal system of weights and measures based on many of the same principles as the metric system, but with units differing in size from those in present use.

Credit for the invention of the system in use today goes to no one person, because it was devised by a number of committees of the Academy of Science in Paris, all working together. Credit for getting the Academy to begin working on it and for the later adoption of the system into law can be pinpointed, though. The prime mover was Charles Maurice de Talleyrand-Périgord, an excommunicated bishop and a political power during the French Revolution. In April 1790 he proposed to the revolutionary National Assembly that the system of weights and measures be overhauled and that a new length unit be adopted based on some unchanging standard found in nature.

The Assembly directed the Academy of Science to investigate the matter. At that time, although the revolution was well under way, the monarchy had not yet been abolished, and Louis XVI still had his head. Louis invited George III of England to send representatives of the Royal Society of London to work with the scientists of the French Academy "to deduce an invariable standard for all the measures and all the weights." George III never accepted the invitation, so the metric system was devised by the French Academy alone.

In October 1790 the first committee report came from the Academy, recommending a decimal base for the new system. In March 1791 the Academy decided that the basic unit of length would be one ten-millionth of the curved line running from the North Pole to the equator through Dunkirk, France, and Barcelona, Spain. That particular meridian was chosen because the area between Dunkirk and Barcelona had already been surveyed, and this would make easier the computation necessary to figure out just what one ten-millionth of the imaginary line was.

Many years later it was discovered that the computation that was eventually made was in error by about two miles. The length of the metre was not changed because of this, but a new, more accurate standard was chosen to define it. That standard and the subsequent one, which superseded it and is in use today, are described in the next chapter.

It was not until 1793 that a name was picked for the new unit. It was called the *metre,* from the Greek word *metron,* which simply meant "a measure."

The metric system did not meet with immediate acceptance. One reason was that it was spawned by the French Revolution, which soon began to appall the world with its radicalism and overuse of the guillotine. It did not help the cause of the metric system when the revolutionists abolished the Academy of Science and guillotined Antoine Laurent Lavoisier, the founder of modern chemistry and one of the important members of the Academy commission that created the metric system. Monarchs became wary of *any* recommended changes in the customary way of doing things which emanated from France, because change seemed synonymous with revolution. No monarch wanted to import any part of the French Revolution to his own country. This feeling wasn't confined to monarchs. Even the leaders of the American Revolution, including Thomas Jefferson, became alienated by the mob violence taking place in France.

The system was not accepted quickly even in France. As has been mentioned, Napoleon abolished compulsory use of the metric system in 1812. But the 1837 act again made it compulsory for all of France, with penalties for not using it. A portion of the act read:

After January 1, 1840, all weights and measures other than those established by law, constituting the decimal metric system, shall be forbidden under penalties. Those possessing weights and measures other than those recognized shall be punished. All denominations of weights and measures other than those recognized are forbidden in public acts, documents and announcements.

That effectively put the metric system in force as the only one used in France after January 1, 1840.

After that date the metric system also began to receive more and more international acceptance. When its use was resumed in France, only Belgium, the Netherlands, and Luxembourg also used it. By the time the United States became party to the Treaty of the Metre in 1875, there were more than twenty metric nations. In 1920 there were thirty-four, and by 1960 there were eighty. Since then the rate of acceptance has been so rapid that today, as has been previously mentioned, of the approximately 145 nations in the world, only the United

States and nine very small and underdeveloped nations have not gone metric.

The fact that organized opposition to the adoption of the metric system in the United States has virtually disappeared does not mean that there are no longer any objectious to the system. There is no organized opposition largely because even the system's opponents realize it is economically impossible for the United States to continue to cling stubbornly to the old system in the face of all the rest of the world going metric. But the system still has critics.

One criticism that sounds valid until you examine it is that with a base of 10 you lose the considerable advantage of the binary system of being able to divide units by 2 repeatedly. Ten can be divided only once, because that gives you 5, which is not divisible by 2.

The exponents of this argument ask how housewives and restaurants are going to slice pies into ten pieces, or how one can fold a piece of paper into ten equal parts. The answer is that no one should try. Such operations do not involve measurement, but merely division. Nothing in the metric system decrees that you can't go right on using such convenient customs as cutting a pie into equal pieces by first slicing it in half, then into quarters, and finally into eighths, or that you will be prohibited from folding a piece of paper into eight equal parts.

An even more common objection to adopting the metric system is that we will not only have to learn an entirely new system, but will have to unlearn the old familiar one. This objection is based on the assumption that we all "know" the present system. But do we really?

The three basic quantities requiring measurement in everyday affairs are length, capacity, and weight. Of course, all of us are familiar enough with the common units of length, so that we could mark off fairly accurate approximations of an inch, a foot, or a yard without a measuring rule. Most of us even know that a rod is 16½ feet and that a statute mile is

1,760 yards or 5,280 feet. But before reading the last chapter,
how many readers knew that a furlong was 40 rods? And how
many remember that a league is 3 miles? (I recall as a child
thinking it was 7.)

Most people have fairly accurate mental pictures of just
how much an ounce by volume is, or a pint, or a quart, or a
gallon. A peck and a bushel are easily visualized too. But how
many readers can define a gill, a barrel, or a hogshead? (For
your information, they are 4 ounces, 31½ gallons, and 63 gallons, respectively.)

You can probably imagine the weight of an ounce, a pound,
and a ton, but can you define a scruple, a pennyweight, or a
dram? (They are, respectively, 20 grains, 24 grains, and 60
grains, but can you visualize mentally the weight of a grain?)

When you include the specialized measurement units used
within certain trades, such as the surveyor's link (7.92 inches)
and the English ell, used by drapers (45 inches), there are
well over a hundred units of measure in the customary system.
Even excluding specialized units, there are more than sixty
separate units for general measurement. Since few people have
more than the vaguest conception of what more than perhaps
two dozen of them mean, it is hardly accurate to say that we
"know" the customary system.

Even some of the common units that we think we know
have different meanings in different places. When you say
mile, are you referring to the statute mile of 5,280 feet, or
the Admiralty mile of 6,080 feet, or the United States nautical
mile of 6,080.20 feet, or the international nautical mile of
6,076.11549 feet? When you say *quart,* do you mean the dry
quart of 67.2 cubic inches, the United States liquid quart of
57.75 cubic inches, or the British quart of 69.355 cubic inches?
When you say *gallon,* do you mean the United States gallon of
231 cubic inches, or the imperial gallon used by our next-
door neighbor, Canada, of 277.418 cubic inches? When you
say *ounce,* do you mean an ounce of liquid or an ounce of
weight? If the latter, do you mean an avoirdupois ounce of

437½ grains or a troy or apothecaries' ounce of 480 grains? When you say *pound*, do you mean the troy pound of 12 ounces, or the avoirdupois pound of 16? When you say *ton*, do you mean the United States short ton of 2,000 pounds, the United States long ton of 2,240 pounds, the nautical ton of 100 cubic feet used to measure cargo capacity, the displacement ton of 35 cubic feet used to reckon a ship's displacement of water, or the freight ton of 40 cubic feet used in shipment by rail?

Even the two specialized units mentioned above, the link and the ell, have different meanings at different times. A surveyor's link is 7.92 inches, but an engineer's link is 12 inches. The English ell is 45 inches, and the Dutch ell, which used to be 27, is now equal to 1 metre.

In the metric system there is no such confusion in terminology. When you say *kilometre*, you always mean exactly 1 000 metres, no matter where you are in the world. When you say *gram*, you always mean the same unit of mass. Unlike the customary ton, which can be either a unit of mass of varying size or a unit of capacity of varying size, the *megagram* (sometimes called a metric ton) is always a unit of mass and is always exactly 1 000 kilograms, no matter what you are weighing or where you are in the world. The *litre* is always the same measure of volume, whether you are measuring ice cream, gasoline, or maple syrup.

As a matter of fact, the metric system eliminates the distinction between liquid and dry measures. The *litre* is simply a special word used to designate a cubic decimetre (a measure of capacity 10 centimetres long, 10 centimetres wide, and 10 centimetres high) when the substance being measured is liquid. The word has come into common use because in metric countries so many liquid commodities in everyday use, such as milk, wine, soda pop, and gasoline, are sold by the cubic decimetre that a less cumbersome name than cubic decimetre seemed necessary, a name that would also indicate that the commodity was liquid. But *any* substance, solid or liquid, can be measured by any of the metric units of capacity. The customary system's confusing double set of values—one for dry measure, an entirely unrelated and different set for liquid measure—is eliminated in the metric system.

The biggest advantage of the metric system over the customary one is its simplicity. Everything is in multiples or divisions of 10.

In the customary system there is neither rhyme nor reason to the numerical progressions used to create multiples or subdivisions of units. The yard is multiplied by 1,760 to make the mile. It is divided by 3 to make the foot. The foot is divided by 12 to make the inch, which in turn is divided into halves, quarters, eighths, and sixteenths. The pound avoirdupois is multiplied by 100 to make the hundredweight, but thanks to Edward I there is also a long hundredweight made by multiplying the pound by 112. The short ton and the long ton are made by multiplying the respective hundredweights by 20. The ounce is arrived at by dividing the pound by 16. Dividing the ounce by 16 gives you the dram, but then you have to divide the dram by 27 11/32 to get the grain. The gallon must be multiplied by 31½ to get the barrel. The barrel is multiplied by 2 to make the hogshead. The gallon is divided by 4 to make the quart, which in turn is divided by 2 to make the pint, which has to be divided by 16 to make the ounce. The multiple of the bushel, the dry barrel, is arrived at by multiplying by 3 9/32. The bushel's subdivision, the peck, is made by dividing by 4. Dividing the peck by 8 makes a dry quart. The dry quart divided by 2 gives you the dry pint.

In the metric system you have to learn only a single progression for *all* units of length, capacity, and mass. For multiples you multiply by 10 and for subdivisions you divide by 10, no matter which quantity you are measuring. This eliminates fractions entirely, because decimals can be used instead.

To illustrate the advantage of this, let us work a simple problem in two ways, first using fractions, then with decimals.

Assume you are at the racetrack, and you picked winners in

the first three races. For your first $2 bet, you receive $13.20; for your second, $6.25; for your third, $8.50. You want to add these three figures together to see how much you have collected in all.

Two methods of addition are available to you. You may use fractions and set down the three figures in this manner:

$$13\tfrac{1}{5} \text{ dollars}$$
$$6\tfrac{1}{4} \text{ dollars}$$
$$8\tfrac{1}{2} \text{ dollars}$$

With fractions of different value, you must convert them all to values of the least common denominator before you can add them. Since the least common denominator in this case is 20, you convert the amounts to those below before adding them:

$$13 \tfrac{4}{20} \text{ dollars}$$
$$6 \tfrac{5}{20} \text{ dollars}$$
$$8\tfrac{10}{20} \text{ dollars}$$
$$\overline{27\tfrac{19}{20} \text{ dollars}}$$

You then convert the 19/20 part of a dollar to cents, and your final answer is $27.95.

You could also have set the figures down like this, though:

$$13.20 \text{ dollars}$$
$$6.25 \text{ dollars}$$
$$8.50 \text{ dollars}$$

Now you have no least common denominator to worry about, and there will be no final fraction that has to be converted back to cents. You simply add the three figures in a single operation and get the same answer of $27.95 about three times as fast.

In this problem you had a choice of methods, but when the

customary system of measurement is used, you have no such choice. Like it or not, you are stuck with fractions. You do have the choice of using the metric system instead of the customary one, though, and eliminating fractions in that way. We will frame another problem to illustrate how much easier the metric system is to use.

A friend offers you five odd lengths of carpeting left over from having his floors covered wall-to-wall. You think there might be enough to carpet a central hallway running the length of your house, but it is going to be very close, so you are going to have to make very accurate measurements.

The hallway measures 31 feet and 3 inches. Carefully measuring the five scraps of carpeting with a yardstick, you get these figures:

$$9 \text{ feet } 3\tfrac{5}{8} \text{ inches}$$
$$7 \text{ feet } 9\tfrac{3}{16} \text{ inches}$$
$$6 \text{ feet } 3\tfrac{1}{2} \text{ inches}$$
$$5 \text{ feet } 9\tfrac{3}{4} \text{ inches}$$
$$2 \text{ feet } 1\tfrac{1}{4} \text{ inches}$$

First you have to convert all the fractions to the lowest common denominator of 16; then you have to reduce all the foot measurements to inches by multiplying by 12; then you have to add the odd inches to those figures to get the totals below:

$$111\tfrac{10}{16} \text{ inches}$$
$$93 \ \tfrac{3}{16} \text{ inches}$$
$$75 \ \tfrac{8}{16} \text{ inches}$$
$$69\tfrac{12}{16} \text{ inches}$$
$$\underline{25 \ \tfrac{4}{16} \text{ inches}}$$
$$375 \ \tfrac{5}{16} \text{ inches}$$

This figure must now be converted back to feet and inches by dividing by 12. That comes out to 31 feet $3\tfrac{5}{16}$ inches, or only $\tfrac{5}{16}$ of an inch more material than you need.

Now let us attack the same problem with a metre stick. The

hallway measures 9 metres 52 centimetres and 5 millimetres. You can write this down in any of three ways you desire, simply by placing the decimal point where you wish. It may be written as 9.525 metres, 952.5 centimetres, or 9 525 milli- metres. You decide to do the problem in terms of metres, and therefore write down 9.525 m.

Now you measure the five scraps of carpeting with your metre stick and get these figures:

$$
\begin{array}{r}
2.835 \text{ m} \\
2.367 \text{ m} \\
1.918 \text{ m} \\
1.772 \text{ m} \\
0.641 \text{ m} \\
\hline
9.533 \text{ m}
\end{array}
$$

Subtracting the 9.525 m length of the hallway from this total, you see that you have 8 millimetres more material than you need. You have obtained the same result with your metre stick as you would with a yardstick in perhaps one third the time and with considerable less chance of error.

Still another advantage of the metric system is that it is so easy to learn. People who grew up in metric countries and later moved to the United States have always had a terrible time learning our customary system because there is no inter- relationship between the various quantities. A woman I know, raised in France, cannot understand why there is no relation- ship between the yard, the gallon, and the pound. Learning that a cubic foot of water weighed 62.4 pounds, she asked querulously, "Why? That relationship doesn't *mean* anything."

She is right, of course. Despite periodic attempts to create some sort of relationship between the quantities of length, volume, and mass in the customary system (such as Henry III's confusing Assize of Bread and Ale), there is no logical relationship between a yard and a gallon, or between a gallon and a pound.

The metric system, on the other hand, is so orderly that one quantity relates to another in a logical and easily understood manner. The keystone of the system is the length unit. The unit of capacity is built upon it, and originally the mass unit was based on volume.

A cubic decimetre, or litre, is 10 centimetres square and 10 high, certainly a more logical arrangement than in the customary system, where the cubic inch has no numerical relationship at all to the gallon.

When the metric system was first devised, the mass unit of the gram was defined as the mass of a cubic centimetre of distilled water at sea level and at a temperature of 4 degrees centigrade (now called Celsius), the temperature at which it had the most density. While that is no longer the SI definition of the gram, for reasons explained in the next chapter, this original relationship is still a handy teaching device, because it points up the step-by-step logic on which the metric system was built. There is no relationship between a gallon and a pound, or a peck and a pound, Henry III notwithstanding. Few users of the customary system have the faintest idea what a gallon of water weighs, but everyone familiar with the metric system knows that a litre of water weighs a kilogram. (Actually, since the international prototype of the kilogram was adopted as the standard of mass, and that eventually proved to weigh slightly more than a litre of water, 1.000 028 litres weighs a kilogram, but that is of little importance outside of a scientific laboratory.)

Probably, the most important advantage of the metric system is that it is the only measurement system ever to approach worldwide adoption. Some 6,500 years after the dawn of civilization, we are finally going to have a universal language of measurement.

Chapter V

SI and the Three Basic Measures

One time a multilingual friend of mine was in a brown study when I broke his train of thought by asking a question. For some moments he gazed at me in confusion before his face finally cleared and he said apologetically, "Excuse my slow return to earth. I was thinking in German."

According to linguistics experts, you have not really mastered a language until you are capable of thinking in that language instead of first forming phrases in your native language, then mentally translating them. A philosophy akin to that is the approach most educators recommend for when the metric system begins to be taught in United States schools. There is even a widely used slogan designed to promote the approach: Think Metric.

In England, where there was no organized program of public education prior to the changeover, the general public has had considerable difficulty adjusting to the metric system. In Australia, where there was a massive educational program prior to conversion, there has been almost no problem at all. Australia used the Think Metric approach.

If the Think Metric system of education is used, schoolchildren will not be taught how to convert customary units to metric ones. They will be taught only the metric system. The customary system will no longer be taught in the schools, and the metric system will be the only one the students will know how to use. The theory is that parents will have to learn the system also in order to keep abreast of their children's progress in school and to be able to help them with their homework.

California Superintendent of Public Instruction Wilson Riles has already announced that Think Metric will be the educational approach in the California schools. Maryland, the only other state at this writing that has set an actual date for beginning to teach metrics in its schools, will probably adopt the same approach, and it is likely that other states will follow these two leaders.

There will continue to be conversion problems, of course, even if nothing but the metric system is taught in the schools. Obviously, the terms in books already published with customary measurements are not going to be magically translated into metric terms. You may find that when you want to build a patio or a barbeque pit or a doghouse, the only plans available have been drawn with the old customary measurements.

Periodically, possibly for the next generation, everyone is going to be confronted with the problem of converting customary units to metric units, but in a Think Metric society, there should be no occasion *ever* for you to have to convert from metric back to customary units. Therefore you will have to learn how to convert from customary to metric units.

You may convert by using conversion tables, conversion factors, or simple electronic calculators available for from $15 to $50. If you want to know what 6 feet 3³⁄₆₄ inches is in metric terms, you may look it up in the conversion tables for length in Appendix B, use the conversion factors in the same appendix, or use a calculator. The answer in any case will come out 1 906.19 mm.

Lacking a calculator, engineers tend to prefer factors as less prone to error, educators tend to prefer tables. But there is by no means unanimity among either group, which is the reason both tables and factors for converting the more common customary units have been included in Appendix B. Tables and factors for converting the other way have been omitted deliberately.

Calculators are widely used in industry, but there is considerable disagreement among educators about permitting their use in classrooms.

Under the International System of Units (SI), an infinite number of units is possible. In our increasingly knowledgeable age, scientists are constantly coming up with new quantities to be measured, and thus new units of measure have to be developed. It is conceivable, for instance, that some future scientist may require a new unit to measure the rate of increase of gravitational pull by the earth on a mass of 1 kilogram approaching the far side of Mars from outer space at half the speed of light. At every General Conference on Weights and Measures, new units are considered.

Some of these units are so esoteric that, quite frankly, I do not have the slightest idea how they are used. No attempt will be made to list all of the present units, even in the Appendix. Those ignored are all in the category called *derived units,* however, which is the category containing all the extremely complicated units, such as the imagined one appearing in the previous paragraph.

Under SI there are at present seven *base units,* forty-three *derived units* in more or less general use, two *supplementary units,* and twenty *accepted units.* But before you let the prospect of having to learn seventy-two new units throw you into a panic, think about how few of the one hundred or so units of the customary system you ever felt it necessary to learn. There is no necessity for you to learn any more SI units than you did customary units. You will have to learn fewer, in fact, because you won't have to memorize a dual set of capacity units, one for liquid measure and one for dry measure.

Unless you enter some highly technical field such as thermodynamics, it is unlikely that ever in your whole life will you have occasion to use such units as the hertz (the unit of frequency), the newton (the unit of force), or the pascal (the unit of pressure) outside of a high-school or college classroom or laboratory in physics.

In the General Motors Company conversion program, workers are taught that most people will have to learn only 6½ new terms: the units *metre, litre* and *gram* and the prefixes *kilo, centi* and *milli*. The ½ term is *Celsius*, which has replaced a metric term most people already know: *centigrade*. Most educators would add the prefix *deci* to this list.

For most everyday purposes, you can get along quite well with the metric system by being familiar with only the units that measure length, volume, mass, time, and temperature. All seventy-two of the units mentioned on page 61 are listed in Appendix A at the back of the book, but except for the seven base units (three of which you will find little use for outside of science classrooms and laboratories) and one important supplementary unit, only a few of them will be mentioned here. Those few will be mentioned only for the purpose of illustrating what qualifies as a unit for each category.

SI Base Units

A base unit is one that is dimensionally independent. That is, it has but one characteristic. The metre, for instance, has only the characteristic of length; the kilogram has only the characteristic of mass. The seven SI base units, the quantities they measure, and their symbols are shown in Table 1.

TABLE 1

Quantity	Name	Symbol
length	metre	m
mass	kilogram	kg
time	second	s
electric current	ampere	A
thermodynamic temperature	kelvin *	K
luminous intensity	candela	cd
amount of substance	mole	mol

* SI also accepts the degree Celsius as equivalent to the kelvin. This is explained in Chapter VI.

Only two of these symbols are capitalized because the SI rule is for all symbols to be in small letters unless they are derived from proper names. The ampere is named after French scientist André Marie Ampère. The kelvin is named after Baron William Thomson Kelvin, Irish-born mathematician and physicist.

SI Derived Units

A derived unit is one with at least two characteristics, such as length and width or length and time. It is formed by multiplying a base unit by itself or by multiplying or dividing a base unit by one or more other units. An example of a derived unit formed by multiplying a base unit by itself is the area unit of the *square metre* (1 metre × 1 metre), the symbol of which is m². An example of a derived unit formed by multiplying a base unit by another unit is the coulomb (1 second × 1 ampere). The symbol for the coulomb is C, and it is the unit of quantity of electricity. An example of a derived unit formed by dividing a base unit by another unit is *metre per second* (1 metre ÷ second). The symbol for metre per second is m/s, and it is the unit of speed.

SI Supplementary Units

The two supplementary units, both purely geometrical, have been called supplementary simply because the General Conference on Weights and Measures has not yet been able to make up its collective mind whether to classify them as base units or as derived units. They are the *radian*, the unit of plane angle, and the *steradian*, the unit of solid angle. Presumably, the CGPM will eventually get around to giving them more definite classifications.

SI Accepted Units

Accepted units are units outside of SI which the CGPM has accepted for use along with SI. Only eight of the twenty have been permanently accepted, and they are placed in this category because they are so widely used that the CGPM pragmatically recognizes that they will continue to be used whether they are accepted or not. Among them are the *hour*, the *degree* (of angle), and the *litre*. (If the last is a surprise because you took it for granted that the litre was part of the metric system, it will be explained a little farther on.)

The other twelve are in general use also, but only in specialized fields. They have been accepted only temporarily and will be reviewed periodically by the CGPM. Among them are such units as the *international nautical mile*, the *knot* (the unit of ship speed), and the *gal* (a unit employed in geophysics to express acceleration due to gravity).

The three basic quantities of dimensional measurement in any measurement system are length, volume, and mass. Please note that I said *basic* quantities, not *base* quantities. Under SI volume is a derived quantity, not a base quantity. It is, however, one of the three quantities most commonly measured in everyday life, and in that sense it is basic.

As has already been explained, these three basic quantities are all interrelated in the metric system. When one is first learning the system, it is helpful to keep these interrelationships in mind, and therefore they will be presented here in their logical order. The unit of length will be explained first; then it will be shown how the unit of volume was derived from that and how the unit of mass was originally derived from volume.

While under the Think Metric approach you are going to be urged to forget the customary system, like withdrawal from an addictive drug, it may be dangerous to attempt a sudden cutoff. Totally discarding the old system before you learn at least the basics of the new one may leave you floundering. During the learning process, the more familiar customary units are handy for comparison purposes, simply to give you some conception of the size of the new units. In order to help you visualize the length of the metre, it will be compared to the yard. So that you may have at least an approximate understanding of how much a kilogram weighs, it will be compared to the pound. But you should understand that these comparisons are made only as a teaching device. Once you have learned to Think Metric—so that the metre brings its own mental image of length into your mind without reference to any other length unit, and you can visualize about how large a kilogram of steak would be without first having to think about the relationship of the kilogram to the pound— you may entirely forget the customary system.

Length

The SI base unit of length, the *metre*, is 39.37 inches long, or a little more than 3⅜ inches longer than a yard. One thousand metres is a kilometre, the length unit which will replace the mile for measurement of long distances. A *kilometre* is 3,280.8 feet, which is about .62 or ⅝ of a mile.

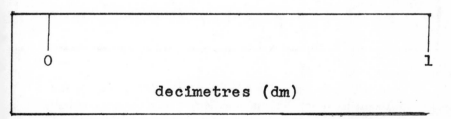

Figure 1

The metre is divided into ten subunits of length called *decimetres*. Figure 1 is the actual size of the first one tenth of a metre stick graduated in decimetres.

The decimetre in turn is divided into ten subunits called *centimetres*. Figure 2 is the actual size of the first one tenth of a metre stick graduated in centimetres.

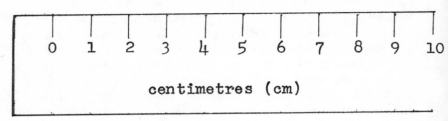

Figure 2

The centimetre is divided into ten subunits called *milli-metres*. Figure 3 is the actual size of the first one tenth of a metre stick graduated in millimetres.

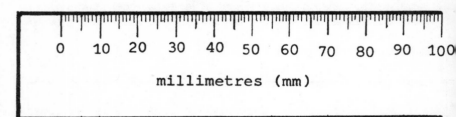

Figure 3

As will be explained a little farther on, there are multiples of the metre other than the kilometre and subdivisions smaller than the millimetre. But as will also be explained, it isn't important for you to learn them. For the moment they will be tabled.

The average mass-produced metre stick sold for home use, like the wooden yardstick you are used to using, is only approximately accurate, because it is not intended for precision work. If it is 1 millimetre shorter or longer than the interna-tional standard, that is hardly important in building a book-case, or even in building a house, because it is an error of only one part in 1,000. But if you used a metre stick with even that small an error to measure the distance from Los Angeles to New York City (2,890 miles), the compounded error would be 4.65 kilometres, or nearly 2.9 miles.

This is only one example of why an international standard of measurement is needed. As another example, consider that even the largest manufacturing companies buy up to 50 percent of the parts they use from other companies that specialize in making certain items. Most big companies, for instance, find it cheaper to buy all their fasteners, such as screws, bolts, and nuts, instead of making them. When a company such as International Harvester turns out a tractor in which the parts from perhaps a dozen different small suppliers are used, the parts had better be made to the same standard, or the tractor won't fit together properly.

The original standard of the metre, as mentioned in Chapter IV, was one ten-millionth of a curved imaginary line running from the North Pole to the equator along the meridian that includes both Dunkirk, France, and Barcelona, Spain. By the time the first General Conference on Weights and Measures met in Paris in 1889, it had been discovered that the computations fixing the length of the metre by that standard were in error.

Meantime the international prototypes of the metre and the kilogram had been constructed under the supervision of the International Bureau of Weights and Measures, as had been provided for in the 1875 Treaty of the Metre. Instead of changing the length of the metre to conform to the actual distance of one ten-millionth of the arc, which would have caused considerable confusion, CGPM left the metre the same length and chose the prototype just constructed as the new standard. The prototypes of both the metre and the kilogram were made of a platinum alloy with 10 percent iridium. The length of the metre bar when it was at 0 degrees Celsius was chosen as the international standard of the metre.

Two things about using a prototype for a standard have

always bothered scientists, though. One is that some unchangeable standard in nature is preferred; the other is that there is always the possibility that the prototype might be destroyed, as were those of England in the fire of 1834.

A search was made for some unchangeable standard in nature which was exactly the same size as the metre prototype. It was finally discovered that 1 650 763.73 wavelengths of the orange light emitted by excited atoms of the element krypton-86 was the exact length of the prototype. At the eleventh CGPM in 1960, this was adopted as the international standard of the metre.

The platinum-iridium prototype has not been discarded, however. It is still kept at the International Bureau of Weights and Measures laboratory, where it is still maintained at a temperature of 0 degrees Celsius. There are also numerous identical copies of it in other parts of the world, including the United States National Bureau of Standards. In practice, these prototypes are still referred to for precise comparison, but the official wavelength standard has relieved the minds of scientists, because it is indestructible, independently reproducible in any laboratory possessing the proper equipment, and universally available.

To recapitulate, the length units you will be most concerned with are listed below:

Unit	Value	Symbol
kilometre	1 000 metres	km
metre	base unit	m
decimetre	0.1 metre	dm
centimetre	0.01 metre	cm
millimetre	0.001 metre	mm

Volume

Figure 4 represents a cubic can, with no top, measuring 10 centimetres long, 10 centimetres wide, and 10 centimetres high. A can of this size holds a cubic decimetre, or a litre.

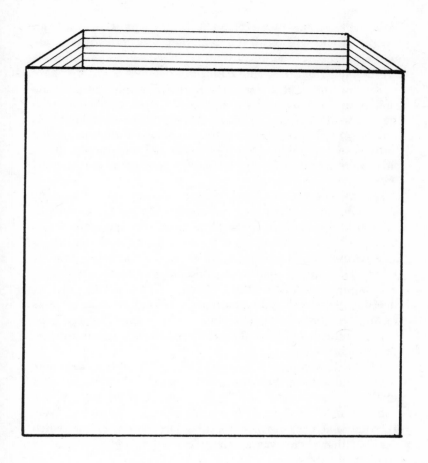

Figure 4

The symbol for the litre is l. It is approximately 33.8 liquid ounces, compared to the 32 ounces in a liquid United States quart.

If you will recall the definition of a base unit—a unit that has but one characteristic—you will see why the litre is not a base unit. It has three characteristics: length, width, and height. If you will recall the definition of a derived unit, you will also see why the litre is not a derived unit. Derived units are derived from base units, and the decimetre is not a base unit. The base unit of length is the metre, and therefore the unit of volume derived from it is the *cubic metre*, not the litre. *Litre* is merely a special name for a cubic decimetre of liquid, and the cubic decimetre is merely a subunit of the derived unit of volume, the cubic metre. This is the reason the litre is not a part of SI, but is merely classified as an accepted unit.

It may not stay in that category forever. SI, despite its strict rules, is a system with a built-in capacity for growth, in that the rules may be adjusted to meet new requirements of the scientific community as they arise. There is no danger of such growth being as haphazard as in the customary system, where the whim of a monarch could create such an anomaly as a 112-pound hundredweight. Changes can occur only by decision of the General Conference on Weights and Measures during its periodic meetings, and such changes must be based on good, solid scientific evidence that they will improve the system.

At the fourteenth CGPM in 1971, for instance, the *mole* was added as a seventh base unit upon the recommendations of the International Union of Pure and Applied Physics, the International Union of Pure and Applied Chemistry, and the International Organization for Standardization, all of which testified that there was a scientific need to define a unit of amount of substance.

It is conceivable that some future CGPM may decide that volume is a single characteristic, and that the litre should become a base unit. In such an event, an international standard for the litre would have to be chosen, but I would not hazard a guess as to what it could possibly be. If and when such a standard is needed, however, you can be sure that scientists will find one somewhere in nature.

So much space has been devoted to the litre, despite its being only an accepted unit of SI, because it is one of the most common units employed in everyday transactions. But it should not be forgotten that the basic (not the base) unit of capacity is the cubic metre.

The cubic metre, like its parent the metre, has both multiples and subdivisions. In actual practice, prefixes are seldom used for multiples of the cubic metre; a number, such as 3 000 cubic metres, is used instead, regardless of the number of cubic metres. But like the metre, the cubic metre is subdivided by tens, and it has the same prefixes for its subunits:

Unit	Value		Symbol
cubic metre	derived unit		m³
cubic decimetre	0.001	cubic metre	dm³
cubic centimetre	0.000 001	cubic metre	cm³
cubic millimetre	0.000 000 001	cubic metre	mm³

Mass

Weight is the attraction of gravity, and it varies according to how far you are from the center of the earth. On top of Mount Everest, which is the highest spot in the world, you would weigh less than you do at sea level; at the bottom of a coal mine you would weigh more. On the moon this book would be so light you could probably toss it easily 100 feet into the air. On the surface of a planet the size of the sun, it would perhaps weigh 50 pounds.

Mass, however, is an unchanging quantity. At sea level, at 4 degrees Celsius, and at standard atmospheric pressure, the weight of a cubic centimetre of distilled water is the same as its mass. But its weight will vary upward or downward according to changes in its environment, whereas its mass will always remain the same, even if it is sent to the moon.

It used to be standard practice in physics textbooks to explain mass as the amount of substance a body possesses, but with the adoption of the mole as the SI base unit for amount of substance, that explanation is too confusing. The mole is the amount of substance in a different sense than that used in the old explanation of what mass is, but that doesn't make it any less confusing.

The actual measure of mass is *inertia*. The mass of a body is determined by the amount of resistance it puts up before a given force applied to it can cause it to begin to move. It is unimportant for you to know the exact scientific definition of mass, however, so long as you understand the distinction between it and weight. While this is far from a scientific definition, it might make the quantity of mass more understandable to you if you thought of it as *unchanging weight*.

Scientists and engineers have always been so bothered by the potential for confusion between the terms *weight* and *mass* that many would like to see the word *weight* eliminated from the vocabulary. At the third CGPM in 1901, the conference declared that the kilogram was the basic unit of mass only, and not of weight also. In industry writers of technical communications are advised to avoid use of the noun *weight* altogether, although the verb *to weigh* is allowed.

Despite the aversion engineers have to the word, in everyday life things continue to be bought and sold by weight. But you should keep in mind that in SI the kilogram and its multiples and subunits are always units of mass, not of weight.

Figure 5 represents the actual size of some representative metric weights of brass.

A complete set of weights would contain many more than these four. A standard set that goes up to a kilogram will contain these weights:

	1 g	30 g
two	2 g	50 g
	5 g	two 100 g
	10 g	300 g
	1 kg	

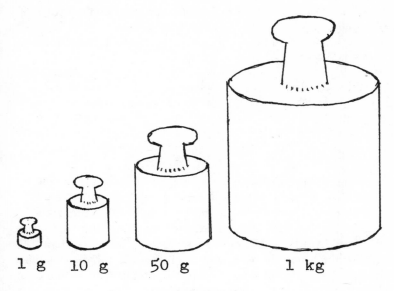

1 g   10 g   50 g   1 kg

Figure 5

There are two 2-gram weights so that combinations can be made to make 3, 4, 7, 8, and 9 grams, thus cutting down on the number of weights necessary. There is no 20-gram weight because 20 grams may be obtained by placing the 30-gram weight on one side of the balance and the 10-gram weight on the other side. There are two 100-gram weights so that combinations can be made to equal 200, 400, and 500 grams. To obtain weights from 600 to 900 grams, you put the kilogram weight on one side of the scale and 400, 300, 200, or 100 grams on the other side.

Weight sets used in laboratories, pharmacies, and other places where delicate measurements are required include subdivisions of the gram. These are flat metal weights of descending denomination which have slightly raised lips on one end so that they may be lifted with tweezers. The very smallest are so tiny and paper thin that a fine laboratory balance, enclosed in glass so that air currents won't disturb it, can weigh a pencil mark on a piece of paper.

The only units of mass you will be likely to encounter in everyday life are these:

Unit	Value	Symbol
kilogram	base unit	kg
gram	0.001 kg	g
decigram	0.1 g	dg
centigram	0.01 g	cg
milligram	0.001 g	mg

A gram is approximately 1/28 ounce. A kilogram is slightly more than 2⅕ pounds avoirdupois.

The reason a multiple of the gram instead of the gram itself is the base unit of mass requires some explanation. When the Paris Academy of Science created the metric system, the gram, which was derived from the mass of a cubic centimetre of water, was intended to be the basic unit of mass. (The term *base unit* did not come into use until 1954, when it was adopted by the tenth CGPM.) By the time of the first CGPM in 1889, scientists realized that this had been a mistake. If mass and volume were to be related, having the gram as the basic unit of mass would logically make the cubic centimetre the basic unit of volume. Not only was the cm^3 (at that time its symbol was cc) too small to be a practical basic unit of volume, but naming it such would put you right back where you started, with another basic unit misnamed, since the cm^3 was a subunit of the cubic metre.

The dilemma could have been solved simply by renaming the kilogram the gram, which would have made the present gram the milligram, but metric nomenclature had been in use long enough by then so that attempting to change it would have been more confusing than having a basic unit with a prefix designating it as a multiple of a unit. It was decided simply to leave well enough alone.

The international standard for the kilogram is still the platinum-iridium prototype adopted by the first CGPM in 1889. There is not the same unease among scientists over the possibility of its being destroyed as there was about the metre bar because it could be reconstructed, since it is based on the mass of water under specific conditions. The prototype is the mass of 1 000.028 cm³ of pure water at maximum density and under standard atmospheric pressure. When it was made, it was meant to be exactly the mass of a cubic decimetre of water, but later, more accurate measurement revealed the overage.

This slight overage, incidentally, is another reason the litre is not part of SI. At the third CGPM in 1901, the litre was actually named the official unit of volume, but instead of merely being defined as a cubic decimetre, it was defined as the volume occupied by "a mass of 1 kilogram of pure water, at its maximum density and at standard atmospheric pressure." This was just the reverse of the original definition of the kilogram, which had been defined as the mass of a cubic decimetre of water. Since that definition had been abrogated in 1889, when the new prototype became the standard for mass, and the official definition of mass was thus no longer in terms of volume, the conference attempted to reestablish the relationship between mass and volume by defining volume in terms of mass.

When it was eventually discovered that the mass of the prototype was the mass of 1 000.028 cm³ of water instead of only 1 000 cm³, a peculiar situation existed—the litre, by official definition, differed from the cubic decimetre by 0.028 cm³.

While this made no difference whatever to the average consumer who was used to buying milk and gasoline by the litre, it bothered scientists. At the twelfth CGPM in 1964, the 1901 definition of the litre was abrogated, and the word *litre* was declared merely a special name for the cubic decimetre. The

litre is, therefore, again officially identical to the cubic decimetre after varying from it by 28/1000 of a cubic centimetre
for sixty-three years.

As with the prototype of the metre bar, identical copies of
the kilogram prototype are stored in various parts of the
world. One is at the United States National Bureau of
Standards.

Chapter VI

SI and Less Commonly Used Measures

You no doubt noticed in the preceding chapter that the prefixes for the units which measure all three of the quantities discussed were identical. This is another advantage of the metric system over the customary system. There is no hint in the word *ounce* that it is 1/16 of its parent unit, the pound. There is no hint in the word *inch* that it is 1/36 of its parent unit, the yard. But from the prefixes used with any of the units of length, volume, or mass in the metric system, you can always tell exactly what multiple or subdivision it is of the basic unit. *Kilo* always means 1 000, whether it is used with the suffix *metre, litre,* or *gram; centi* always means a 1/100, and *milli* always means a 1/1 000.

Only four prefixes are commonly used in everyday transactions, but under the International System of Units there are twelve. There is no point in your memorizing the seldom-used eight, but you should at least be aware of their existence, so that if you do run across one, you can look up its meaning.

All the prefixes used to designate multiples in SI are Greek in origin, and all the prefixes used to designate subdivisions are derived from Latin. There are six of each.

PREFIXES FOR MULTIPLES

Multiplying Factor	Prefix	Symbol
10	deka	da
100	hecto	h
1 000	kilo	k
1 000 000	mega	M
1 000 000 000	giga	G
1 000 000 000 000	tera	T

PREFIXES FOR SUBDIVISIONS

Multiplying Factor	Prefix	Symbol
0.1	deci	d
0.01	centi	c
0.001	milli	m
0.000 001	micro	μ
0.000 000 001	nano	n
0.000 000 000 001	pico	p

Like many of the more complicated derived units, most of these prefixes are used only in the laboratory. The only commonly used multiple prefix is kilo; the only three prefixes in common use for subdivisions are deci, centi, and milli.

You will rarely, if ever, hear anyone say dekametre or hectometre in describing distance. Common practice is merely to say 10 metres or 100 metres. Nor are you likely to see megametre used, even on a map of the world. A distance such as 2 megametres is almost always written simply as 2 000 kilometres. Common usage is to say 10 or 100 grams instead of dekagram or hectogram. Megagram is used in certain heavy industries, but you are unlikely to hear it in everyday use. You may hear 1 000 kilograms referred to as a metric ton.

If you drove into a service station and ordered 4 dekalitres of gasoline, the average attendant, even in France, would probably think you were trying to show off. You would get much quicker service at most stations by ordering 40 litres.

When subdivisions of units smaller than 1/1 000 are used,

the practice outside of laboratories is simply to express them in decimals. Instead of saying micrometre, you would ordinarily say 1/1 000 of a millimetre, and instead of saying nanogram you would ordinarily say 1/1 000 000 of a milligram.

So you should remember that there *are* prefixes for multiples and subdivisions of units other than the four in common use. If you go to the trouble of memorizing all of them, however, in all probability you will have forgotten the one you want to use the first time an occasion arises to use it.

In the preceding chapter we covered the three quantities of length, volume, and mass, but only length and mass are base quantities. That leaves five base quantities we have not yet touched upon. Unless you plan to enter some scientific field, you will find practical use for only two of these: time and temperature. The other three—electric current, amount of substance, and luminous intensity—will be defined in order to make the coverage of base quantities and base units complete, but only very briefly.

Time

Time is a quantity which we continue to measure repeatedly as long as we are awake. We measure it in order to decide when to get up in the morning, when to get to school or work, when to eat, when to go to bed, and when to perform countless other acts.

Although the time of day at any given moment varies in twenty-four different time zones around the world, the entire world is on the same time *system*. It is necessary for the system to be coordinated accurately on a worldwide basis at all times, so that those in Shanghai, for example, can know the exact time in New York City and vice versa. Airplane schedules depend on this, long-distance operators depend on it, and countless other international transactions depend on it.

A factor in coordinating time throughout the world is the scientist's ability to measure it accurately, and that ability has steadily increased over the centuries. We won't bother to go clear back to the invention of the sundial, but will start the discussion with the relatively modern period when scientists first began to attempt to define and measure the second.

The original method of defining the second, before the atomic age ushered in devices for measuring time to an accuracy never dreamed of by preatomic-age scientists, was in relation to the 24-hour day. The day was defined as the time required for the earth to make one complete turn on its axis. It was divided into 24 equal periods called hours, which in turn were divided into 60 equal periods called minutes, which in turn were divided into 60 equal periods called seconds. That made the second 1/86 400 of a day.

That seemed simple enough to everyone concerned, including scientists, until astronomical measurements disclosed that the earth didn't always take the same amount of time to make one revolution on its axis. The earth's orbit around the sun, which it completes every 365 and a fraction days, is not an exact circle, but an ellipse, which means that the earth is at varying distances from the sun at different times of the year. The time it takes the earth to make one complete revolution on its axis varies according to its distance from the sun.

To compensate for this, astronomers averaged together the 365 and a fraction revolutions the earth made during each complete orbit of the sun, and called this average the *mean solar day*.

Three different methods were used to compute the length of a year, however, and they gave three different answers. When scientists measured the year from the time the middle of the sun crossed the equator until it again crossed the equator 365 times later, the year came out to be 365 days, 5 hours, 48 minutes, and 45.5 seconds. Scientists called this the *tropical year*. When they measured the year from the sun's eclipse of a certain star until its next eclipse of the same star, the year came out to 365 days, 6 hours, 9 minutes, and 9.54 seconds. They called this the *sidereal year*. When they measured the time that elapsed between the earth's leaving a particular point in its orbit until it arrived back at the same point, the year came out to be 365 days, 6 hours, 13 minutes, and 53.1 seconds. They called this the *anomalistic year*.

At the eleventh CGPM in 1956, the second was defined as the fraction of the tropical year of 1/31 556 925.974 7. Experiments were already under way at that time, though, to define the second even more accurately in terms of atomic radiation. By the time of the thirteenth CGPM in 1967, those experiments had been completed, and the second was redefined.

The SI second is now defined as the time it takes the cesium-133 atom to vibrate 9 192 631 770 times. It seems unlikely that a more accurate measurement will be discovered in the near future.

While the new definition of the second was exciting news in the scientific community, it excited only a yawn from the average commuter, who went right on setting his alarm clock by the six o'clock newscast. The fact is that there is nothing *new* you have to learn about time measurement under the International System of Units. The same time divisions are used as under the customary system. The only differences, neither of which affects the everyday use of time measurement, are that in the SI system the second is defined more accurately than ever before in scientific history, and that the second is the *only* time unit actually a part of SI. The other units in use are merely accepted units.

This is because in laboratory work the multiples and subdivisions of the second are in tens, just as they are for other SI units, whereas in the customary system the multiples are by sixties. In the laboratory time is measured in such denominations as 1 000 seconds, not 16 minutes and 40 seconds. The customary multiples have been in worldwide use for everyday purposes for so many generations, though, that it would be impractical to attempt to establish new units based on the ten system.

Below are the time units you will continue to use under SI. Keep in mind that all but the base unit are merely accepted units used with SI.

Unit	Value	Symbol
second	base unit	s
minute	60 s	min
hour	3 600 s	h
day	86 400 s	d

Temperature

Heat is a form of energy. There is no such quantity as *cold;* that is merely a word used to designate the absence of heat. All objects, at all temperatures above what is known as *absolute zero,* contain at least some heat. If it contains a small enough amount, we say that an object is "cold," but actually all that means is that it contains less heat than our fingers, so that when we touch it, heat is drawn away from our fingers into the object, creating a sensation of coolness.

Heat is caused by the movement of the molecules and atoms of which matter is constructed. This movement may be stimulated by a chemical reaction or by a transference of energy. When you burn coal, this is a chemical reaction; the carbon in the coal is combining with the oxygen in the air to form carbon dioxide and carbon monoxide. When you hold a poker in a fire, heat energy is transferred to the poker. In both instances the atoms and molecules of the matter concerned are stimulated to move more rapidly, causing heat.

The faster the motion of atomic particles is, the hotter the object containing them becomes. The slower the motion is, the colder it becomes. All atomic movement stops at absolute zero, which is –273.16° C or –459.69° F.

Absolute zero has never actually been obtained, but scientists have been able to get within less than 1/100° Celsius of it.

Temperature is the measure of the *intensity* of heat in a body. It has nothing to do with the *amount* of heat, because that depends on factors such as mass and the capacity of different substances to generate heat. Two bodies may contain vastly different amounts of heat, but if the intensity of their heat is identical, they are at the same temperature.

Human bodies are highly sensitive to temperature. At 70° F a room will feel comfortable; at 68° F we are apt to complain of a chill. A mother can tell when her baby has a fever simply by feeling its cheek. I once knew an elderly woman who had cooked all her life on an old-fashioned coal range. When her husband finally bought her a modern gas range with a numbered temperature control for the oven, she used to stick one hand in the oven and use her other hand to adjust the control, without looking at it, until the oven temperature felt just right to her hand. When I checked the temperature control on her oven, I would find it set to within a degree or two of what the cookbook specified as the proper temperature for whatever she was baking.

Primitive attempts to measure temperature were based on natural phenomena. When the air reached a certain coolness, people's breath became visible. When the air became a certain amount colder, water froze. When an iron rod reached a certain heat level, it turned red. The Australian aborigines, who sleep with their dogs instead of using blankets, rate the temperature of the night air by how many dogs are required to keep them warm. A nippy night is a two-dog night; one that is bitterly cold is a six-dog night.

Attempts to devise a more accurate system of measuring temperature go back at least as far as ancient Greece. It was the sixteenth century before anyone succeeded, though, and that was hardly an unqualified success. Italian scientist Galileo Galilei (usually referred to by only his first name) is the first known person to have created a thermometer. It was an extremely inaccurate one. The Greeks had discovered that air expands when heated, and Galileo used that principle.

His thermometer, which he constructed in 1593, consisted of a glass tube that was open on one end and had a bulb on the other. The open end of the tube was calibrated in equal divisions. Galileo sucked air from the tube; then he immersed the open end in a vessel of colored water, and the water was drawn up into the tube for a certain length. When a heated object was pressed against the bulb or a flame was held to it, the expanding air forced the water in the tube back downward. Theoretically, the level that the water was forced to registered the temperature of the object or flame.

I say theoretically because even when the same sort of flame was used, the same temperature was seldom registered twice in a row. One reason for this was that Galileo failed to take atmospheric pressure into account, and that also caused changes in the height of the column of water. But the main reason was that an air thermometer is simply too clumsy a device for accurate measurement.

In 1641 Ferdinand II de Medici, Grand Duke of Tuscany, made a scaled thermometer consisting of colored alcohol in a sealed glass tube. Alcohol expands when heated and contracts when cold. It has a further advantage over water in that it does not freeze until a very low temperature is reached, and therefore temperatures below the freezing point of water could be measured.

Ferdinand's scale was merely one of arbitrary numbers which had no relationship to any constant temperature found in nature. In other words, it had no standard. Twenty years later another scientist attempted to set a standard for Ferdinand's thermometer by making a new scale for it based on two reference points, which he considered "fixed" temperatures in nature. One was the temperature of snow, the other of "midsummer heat." Both of these turned out to be variables instead of constants, but the idea of a temperature scale having a cold reference point and a hot reference point, both based on some natural constant, was a good one that was adopted by later scientists.

In 1664 English mathematician and inventor Robert Hooke proposed the freezing point of water as one of the fixed temperatures, but it was another thirty years before it occurred to anyone to make the boiling point of water the other one. That suggestion came from a scientist named Renaldini.

In 1709 a German instrument maker named Gabriel Daniel Fahrenheit made an alcohol thermometer and began to experiment with it. He noted that melting ice always gave the same temperature reading. He suspected that boiling water had the same constancy of temperature, but he couldn't use his alcohol thermometer to test that because alcohol boiled before water did.

It occurred to him that mercury might be a much more suitable liquid than alcohol for measuring temperature. It remained liquid through a long range of temperatures. (Mercury doesn't freeze until 38° below zero on the Fahrenheit scale, or boil until 674° above, although of course Fahrenheit didn't know this because he had not yet invented the scale.) Mercury expanded when heated and contracted when cooled. It was opaque and therefore could be seen easily in a narrow column, and it did not wet the inside of the glass tube in which it was enclosed, as alcohol did. In 1714 Fahrenheit built the first mercury thermometer.

He wanted to calibrate his temperature scale according to two fixed references, which could be duplicated in any laboratory, so that other thermometers could be calibrated identically without having to be compared to his thermometer. But for the purpose of experimenting in order to decide on those reference points, he merely marked off his thermometer in arbitrary divisions.

Repeated experiments with both alcohol and mercury thermometers revealed to him that the temperature of a mixture of ice and salt was always the same. That was the lowest temperature Fahrenheit was able to produce with the laboratory facilities available to him. He called it zero.

Some years earlier Sir Isaac Newton had published the results of some experiments with oil-filled thermometers. The thermometers were not very successful, but Fahrenheit decided to adopt two principles from Newton's report. One was that the body temperature of all persons in good health was identical; the other was that a temperature scale of twelve divisions was logical because it conformed to the twelve inches in the foot. Fahrenheit, therefore, set the twelfth mark above zero on his scale as the temperature of the human body.

He soon realized that these divisions were too large for an accurate measurement of temperatures, though, so he divided each unit in half to give a scale of twenty-four divisions. When he found that that was not a fine enough scale either, he divided again, into forty-eight units; then he finally decided to make one last division of all units, and ended up with a scale of ninety-six degrees. This last number, according to Fahrenheit's calculations, was the unchanging temperature of the human body.

Apparently, Gabriel Fahrenheit was one of those people whose normal temperature was below that of the average person's, because it eventually developed that on the Fahrenheit scale the average body temperature was 98.6°. It further developed that not only did body temperature vary from person to person, it even varied within individuals during different periods of each 24 hours, in some cases by as much as two degrees. So the second constant on which Fahrenheit based his scale turned out not to be a constant after all.

The scale was already set, though, so he left it as it was. Using that scale, he eventually determined that the freezing point of water was 32°, and the boiling point was 212°. Those two values are still the same today on the Fahrenheit scale.

In 1731 French physicist René Réaumur constructed a thermometer called the Réaumur thermometer, which is still in use in a few places in Europe. It was graduated so that 0° was the freezing point of water and 80° was the boiling point of water.

In 1742, only two years before he died at the relatively early age of forty-three, Swedish astronomer Anders Celsius devised a temperature scale based on Réaumur's principle that the freezing and boiling points of water should be the two standard reference points, but Celsius had 100 instead of only 80 divisions between those two points. His original proposal was the reverse of the scale in present use, however. On his scale water boiled at 0° and froze at 100°. Shortly before his death, another scientist suggested that the scale be inverted, so that 0° was the freezing point of water and 100° its boiling point. Celsius accepted the suggestion, and that became the scale for what came to be called the centigrade thermometer.

In recent years the name of the scale has been changed from centigrade to Celsius, after the name of its inventor.

Figure 6 shows the relationship between the Celsius and the Fahrenheit scales.

Temperature is one area in which the customary system will probably continue to be used alongside the metric system for many years. This is because thermometers are simply too expensive to throw away. Even if all doctors discard their Fahrenheit clinical thermometers the moment we go metric and buy new ones scaled to Celsius, there will still be millions of Fahrenheit thermometers in home medicine cabinets, which frugal housewives will refuse to throw away. Even if the news media begin using Celsius temperatures in all weather reports, millions of thermometers on private homes will continue to record the temperature outside in Fahrenheit. And, of course, the temperature gauge on every thermostatically controlled home furnace in the country is scaled to Fahrenheit.

In this instance, therefore, it is necessary for you to be able to convert both from the customary system to metric and vice versa. Fortunately, conversion is quite simple.

To change degrees Fahrenheit to degrees Celsius, subtract 32 from the Fahrenheit temperature and multiply by 5/9. To convert from Celsius to Fahrenheit, multiply the Celsius temperature by 9/5 and add 32.

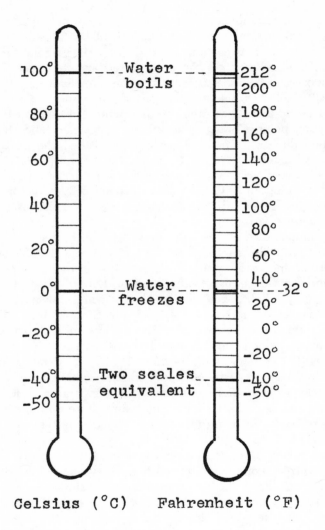

Celsius (°C)   Fahrenheit (°F)

Figure 6

The two temperatures the average person has occasion to measure most often are air temperature and body temperature. Everyone knows that the air temperature generally regarded as ideal for comfort is 70° F, and that normal body temperature is 98.6° F. It will help you to begin to Think Metric as far as temperature is concerned to fix the Celsius equivalents of those temperatures in your mind.

The temperature 70° F does not convert exactly, but it is close to 21° C, so that this reading is the one at which you would set a Celsius thermostat for comfortable room temperature. The temperature 98.6° F converts to 37° C exactly.

You have probably begun to wonder why, after all this discussion of temperature, the SI base unit of the kelvin still has not been mentioned. The main reason is that its use is pretty well confined to the laboratory, and for all ordinary purposes you will be using the degree Celsius instead of the kelvin. As units of measure the two are identical in amount, but they are applied in different ways.

Lord Kelvin, after whom the SI unit of thermodynamic temperature was named, was the inventor of the *absolute temperature scale*. Contrary to the Fahrenheit and the Celsius scale, this scale has no numbers on it less than zero, because it starts at absolute zero. That, as mentioned earlier, is –273.16° C or –459.69° F. The divisions of the Kelvin scale are identical in size to those of the Celsius scale, but because the numbering starts at absolute zero, instead of at the freezing point of water, the number of kelvins is always 273.16 higher than degrees Celsius. Thus on the Kelvin scale, the freezing point of water is 273.16 kelvins instead of 0° C, and the boiling point is 373.16 kelvins instead of 100° C.

The kelvin is the unit of *thermodynamic temperature*. Thermodynamics is the science that deals with the relationship between heat and work. It is too complex a subject for discussion here, since this is a book on metrics, not a textbook on physics. Unless you contemplate a career in thermodynamics, all you really have to know about the kelvin is that the number of kelvins is always 273.16 higher than degrees Celsius. Incidentally, correct usage of the word *kelvin* is to use it alone. You do not say 100 degrees kelvin, but simply 100 kelvins.

The *interval* between 2 kelvins and between 2 degrees Celsius is identical, and therefore the kelvin and the degree Celsius are identical units of measure. The distinction between them is merely that the kelvin is used in the thermodynamics laboratory, the degree Celsius in everyday life. Both are integral parts of the International System of Units, but under ordinary circumstances you will use only degrees Celsius.

In the general activities of day-to-day living, you will have little, if any, occasion to use the remaining three base SI units. Furthermore, being units mainly employed in the laboratory, they are already in general use in the United States and therefore can hardly be considered "new" units. But just to be comprehensive, we will touch on all three briefly.

Electric Current

The ampere, the symbol of which is A, is the base unit from which all other electrical units such as the volt, watt, farad, and ohm are derived. It is the unit of intensity of current, and used to have the relatively simple definition that it was "the amount of current that would be produced by an electromotive force of 1 volt acting through a resistance of 1 ohm."

The difficulty with that definition, as far as the General Conference on Weights and Measures was concerned, was that it defined the ampere in terms of two of its derived units, which is something like defining the metre as the cube root of the cubic metre. The ninth CGPM in 1948 overcame that objection, but simultaneously complicated the definition by declaring:

The ampere is that constant current which, if maintained in two straight parallel conductors of infinite length, of negligible circular cross section, and placed one metre apart in vacuum, would produce between these conductors a force equal to 2×10^{-7} newton per metre of length.

This definition is based on the fact that when current flows through a wire, a magnetic field is produced around it. Since the strength of that field is proportional to the current, measuring the field gives an accurate measurement of the current. By referring to the precise conditions under which two wires that are 1 metre apart will produce a field of specific strength when 1 ampere of current is sent through them, CGPM accurately defined the ampere without defining it in terms of its own derived units.

The various derived units of the ampere may be found in Appendix A at the back of the book.

Amount of Substance

Amount of substance is an entirely different quantity from mass. You will recall that mass is related to the amount of force necessary to overcome a body's resistance, but that the common classroom explanation of it in the past was that it meant the "amount of substance." *Amount of substance* under SI has an entirely different meaning from this. The difference is similar to the difference between heat and temperature. *Heat* refers to the total quantity of energy of a specific kind contained in a body. *Temperature* refers to the intensity of the heat. Similarly, *mass* refers to the total quantity of substance in a body, and *amount of substance* refers to the concentration of substance.

The official SI definition of the mole, the symbol of which is mol, is that it is "the amount of substance of a system which contains as many elementary entities as there are atoms in 0.012 kilogram of carbon-12."

The word *system* is used to mean a specific amount of a chemical element or compound. The term *elementary entities* means atoms, molecules, ions, electrons, and other minute particles that make up matter. If a system contains *twice* as many elementary entities as there are atoms in 0.012 kg of carbon-12, the amount of its substance is 2 moles.

This explanation is somewhat simplistic, because actually the mole may be used as a measuring unit in relationship to only one of the elementary entities of a system, or in relationship to specified combinations of them, or in relationship to all of them.

You will not encounter the mole outside of a chemistry or physics laboratory.

Luminous Intensity

The *candela,* the symbol of which is cd, measures the intensity of light. It is based on the amount of light striking a specific area of totally black surface when the light rays are directed straight at the surface, and not at a slant, under specified conditions. The actual SI definition is:

The candela is the luminous intensity, in the perpendicular direction, of a surface of 1/600 000 square metre of a blackbody at the temperature of freezing platinum under a pressure of 101 325 newtons per square metre.

You are not likely to encounter the candela either outside of a laboratory.

Chapter VII

Effects on Business and Industry

The news that the United States is going metric has probably dismayed a lot of county recorders across the nation, because they wonder if all property records and maps will have to be redone in metric terms. It has probably upset many architects, construction engineers, and builders also. Will all future buildings have to be constructed to brand-new specifications? What about lumber sizes? If you want to make repairs on a home built to the old specifications, will you be able to find two-by-fours or 2-inch joists or any of the old, familiar board sizes you need?

It seems unlikely that serious problems will develop in either of these areas. Conversion is voluntary, except for agencies of the federal government. No one is going to force any governmental level lower than the federal to convert, and no one is going to force any privately owned business to do it.

Industry is converting largely because of economic necessity; it has to in order to continue to compete in the world market. But the recording of deeds is not a competitive business. No dire economic consequences are going to result if the County of Los Angeles goes right on using yards, feet, and inches instead of metric units in its Department of Maps and in its recorder's office.

A customer looking for a car built to metric standards can go to another manufacturer if the first company he tries makes cars only to customary standards. But what can he do if he doesn't like having the purchase of property he has just bought in downtown Los Angeles recorded in terms of feet? Have it recorded in San Diego County?

Obviously, local governments will move very cautiously, if at all, in converting their property records to metric standards. Those that do decide to convert will probably do it on a gradual basis, entering records in *both* metric terms and customary terms, and only when property changes hands. It might take two hundred years before conversion is complete, but there is really no big hurry. Property is something that requires measurement perhaps once in a lifetime or even only once in several lifetimes, and you can live just as cozily on a piece of land whose deed describes it in terms of feet as you can on a piece of land described in terms of metres.

Architects and those in the construction industry will probably have to convert eventually, but that doesn't necessarily impose any enormous problems. As far as architects are concerned, conversion should involve little more than their brushing up on the metric system and buying a new set of rules and measuring instruments. In the building industry the major changes could be merely in terminology.

The so-called two-by-four board has never been that measurement. It runs about 1⅝ inches by 3⅝ inches. Calling it a four-by-nine (centimetres) won't describe it exactly, but the description will be considerably more accurate than the present one. Other lumber sizes can be similarly redesignated, with a resulting increase in descriptive accuracy in virtually all cases.

This is not to say that conversion to metric is not going to require any changes other than terminology in the building industry. Both pipe sizes and the pitch of threads at pipe joints differ in the metric system. The sizes of screws and the pitches of their threads differ also. Even wire sizes are different. But overall the necessary changes and adjustments should be relatively minimal.

The greatest impact of the conversion to metrics in the United States, from the standpoint both of initial costs and of eventual benefits, will undoubtedly be on the manufacturing industry. Estimates of the total national cost of conversion— some made by top industrialists, some made by economists, some the result of surveys of key industries—have been published periodically over the past ten years. They vary all the way from fifty to a hundred billion dollars.

The Metric Study Act of 1968, directing the Secretary of Commerce to arrange for a three-year study of the entire metric subject, specified among other things that the study should include an evaluation of both the costs and the benefits. As far as costs are concerned, and findings of the National Bureau of Standards, which as part of the Department of Commerce was designated by the Secretary of Commerce to make the study, were that previous estimates of costs must have been picked out of the air. The conclusion was that there was no way to estimate the total cost.

One reason for this is that before a total estimated cost of any validity could be arrived at, it was necessary to get reliable estimates of the cost of conversion from individual manufacturers. When the Bureau of Standards surveyed industry, the estimates from companies of similar size, manufacturing similar products, varied so widely that it was obvious they were mostly wild guesses. When the Bureau questioned English manufacturers who had already converted to metric, it discovered that even they couldn't come up with any realistic estimates of what conversion had cost them.

Direct costs are easy enough to compute, of course. The cost of a new machine, or the cost of modifying an old one to metric standard, plus the cost of retraining its operator add up to the direct cost. But the increased rate of production because metric computations can be made faster, the reduction in the amount of scrap because workers are less prone to error using the metric system than using the customary system, and the increase in potential market for the finished product have to be balanced against the direct cost in order to get the true cost.

Companies that have already converted to metric have also discovered a bonus in side effects. Since a change was taking place anyway, people decided to review the whole manufacturing process, and along with conversion, many new time-saving techniques have been adopted, which probably would not have been adopted without conversion. Dialogue like the following has become common in machine shops where conversion is taking place:

Drill-press operator: As long as we're converting anyway, why don't we drill four plates at a time instead of only two?
Foreman: Okay, but why didn't you think of that with the old machine?
Drill-press operator: I thought you wanted me to drill only two at a time.

Another factor in computing costs is that the initial outlay is made only once, but the benefits go on indefinitely. So many intangibles are involved that the actual cost of conversion in dollars and cents will probably never be known.

One thing that became clear from the Bureau of Standards' study was that whatever the cost of conversion, it will be less if it takes place according to some orderly plan, coordinated on a national level, than if the nation merely continues its haphazard drift toward metrication. Jeffrey Odom, one of the bureau people who was involved in the metric study, points out that the question of cost is to some extent academic, because realistically industry does not have the alternative of *not* spending the money. We *are* going metric at ever increasing acceleration, so the cost is inevitable and unavoidable. To be sure, only about 25 percent of American industry is now considering metrication seriously enough to have begun making plans to convert, but before too many years have passed the snowball effect of more and more industries converting is going to leave those still on the inch system only two choices: convert or go out of business.

Not only will the inability to compete with a growing consumer market demanding metric products confront unconverted companies with this choice, but also the economic interrelationship between manufacturers in America will act to force conversion on those reluctant to make the change. As has been mentioned, even the largest manufacturers buy about 50 percent of the parts used in the manufacture of their products from other companies. This is such a common practice that there is a special name in the world of manufacture for these suppliers: They are called vendors. When giants such as General Motors, the Ford Motor Company, and IBM finish converting, their vendors will have no choice but to convert also if they want to stay in business.

This interrelationship between different companies is one of the reasons the conversion program is going to have to be coordinated on a national level. A company such as General Motors, which turns out finished products for a consumer market, can contain the problems of conversion pretty well within its own company. It can set whatever timetable for conversion it pleases, based solely on its estimate of the available market for vehicles made to metric standards and its own internal problems.

But consider the plight of its poor vendors. Specifically, let's consider the main vendor from which it purchases the various electric switches used in General Motors vehicles. Let's assume that the Ford Motor Company is another major customer of the vendor, and that between them G.M. and Ford account for half of the switch manufacturer's business. None of the remaining 50 percent of the vendor's customers have any immediate plans for converting to metric.

Obviously, this vendor has a more complicated conversion problem than either G.M. or Ford. The company's timetable for conversion is not merely a matter of internal decision, as it was for its two huge customers, but is forced on the vendor by these two customers' timetables. The company can't even make complete conversion, because half of its customers still require switches made to inch-system standards.

One entire industry may be in this plight before long: the fastener industry. The fastener industry is unique in that nearly all of its business is vending to other companies instead of to the consumer market.

Fasteners are nuts, bolts, screws, rivets, cotter pins, washers, and a variety of other devices used to hold the components of products together. When the various sizes of all fasteners are included, there are nearly two million different kinds that are manufactured. The annual production in the United States is estimated to be more than 200 billion separate fasteners, at a sales figure of about two billion dollars. The average automobile contains about 3,500 fasteners of various types. These are turned out by 660 manufacturing plants with a total of 67,000 employees.

While a few inch-system fasteners are interchangeable with metric fasteners, in most cases they are not. For one thing, the threads of screws and bolts have different pitches in the two systems, so that even when they are the same diameter, they won't screw into the same hole.

The entire fastener industry is at the mercy of its users insofar as metrication is concerned, because it must furnish fasteners to the specifications required. Fastener companys' conversion to metrics, therefore, has to be geared to conversion by the companies it supplies. The industry totally lacks the freedom of decision to convert or not convert which manufacturers such as G.M. or Ford can exercise.

While the Bureau of Standards' metric study report stressed that conversion should be voluntary insofar as legislation is concerned, coordination of planning on a federal level could help resolve problems such as this. While industries planning conversion generally inform their vendors of their timetables, these are usually flexible. Manufacturers who move faster than originally planned may find their vendors haven't kept pace;

those who move slower could leave their vendors with inventories of unsalable parts. A government-coordinated program of collecting information on what and where surplus metric parts are available to manufacturers who are converting would ease this problem for all concerned.

In addition to the Ford Motor Company, IBM, General Motors, and International Harvester, some other companies that are either studying the advantages of going metric or are in some stage of metrication, are Honeywell, Inc., the Regal-Beloit Corporation, ITT Gilfillan, Caterpillar Tractor, Deere and Company, and the Chrysler Corporation. In doing research for this book, I have either read or personally listened to the views of the executives most concerned with the metrication programs in all ten of these companies. The following observations represent a composite of their views. In some cases observations represent only a single viewpoint, but in many cases they represent either unanimous or near unanimous opinion.

There is general agreement with the statement by Lowell Foster, who is in charge of the metrication program for Honeywell, that "it won't cost you what you think it will." By this Foster means that even the direct costs, before eventual savings and other benefits have been balanced against them, will generally be less than anticipated. Estimates of the probable direct costs of conversion have in virtually every case been too high. There are a number of reasons for this. One is that companies are often unaware of the potential for converting their present machines instead of replacing them.

For instance, ITT Gilfillan of Van Nuys, California, discovered that some of its most expensive machines, made in Italy, had been made to metric standards and then converted to inch standards for sale in America. Conversion kits were available from the Italian manufacturer, so that the machines could be converted back to metric at a fraction of the cost of replacement.

All ten companies discovered that many inch-system machines could be converted simply by replacing minor parts, and that some could be converted without even replacing any parts, but merely by making adjustments. A drill press, for example, functions whether the holes it drills are in metric sizes or customary sizes. Conversion merely involves the use of different-sized drills. All that is required to convert a machine lathe from an inch-system machine to a metric machine is a new length scale, costing a few dollars, and metric-gauge calipers for measuring the diameters of lathe turnings instead of the inch-gauge set of calipers the lathe operator has been using.

Kenyon Taylor, president of Regal-Beloit Corporation (formerly known as the Beloit Tool Corporation), estimates that as a rule of thumb the average manufacturing plant converting to metrics will find that out of every hundred machine tools, twenty-five would be ready for replacement sometime in the near future anyway; twenty-five will require no alterations; twenty-five will need only small adjustments in order to be converted; twenty-four will require the replacement of dials, scales, and other minor parts in cost ranges of under five hundred dollars; and one major piece of equipment will require either replacement or substantial and costly conversion. Admittedly, there will be wide variations from this rule of thumb, depending on the nature of the manufacturing plant, but it is probably a fair evaluation of what American industry as a whole will run into in converting.

As a result of their experience, the management of the four major companies farthest along the road to conversion at the present time—Ford, G.M., IBM, and International Harvester—generally agree that companies would be wise to follow certain principles when converting to metrics:

(1) Metrication should take place only according to a carefully worked out plan, after a thorough study of all phases of the problem, and according to a timetable consisting of at least several years.

(2) The period of conversion should be as short as is

economically feasible, but care should be taken not to rush things. Each phase of conversion must be carefully worked out. For instance, sufficient advance notice must be given to ven dors to prepare them for the changeover, and assurance must be obtained from them that they will be able to meet the new needs on specified dates. An extreme example of what can happen when planning isn't thorough is the situation that one of America's major companies ran into when attempting to make an addition to one of its foreign plants. The addition was planned to metric specifications, but the company neglected to make sure that all metric components were available before beginning construction. When it was belatedly discovered that some components were not available, construction was held up a full two years.

Top management of the company that made this boner is sensitive about it, and I promised the executive who told the story not to divulge the company's name. The incident illustrates that even our industrial giants are not always as efficient as their public relations people suggest, however

(3) The conversion timetable must be flexible. A rigid timetable is likely to add costs by forcing the company to make conversions at inappropriate times just to keep to the schedule. The IBM conversion process, which has now been under way about two years, is scheduled to take from six to ten years. IBM management hopes complete conversion can be accomplished in six years, but flexibility is provided so that planning can be modified to adjust to any unexpected problems that arise.

(4) Some specific person or group within the company should be appointed to coordinate the metrication program. Companies have found that conversion goes much more smoothly when there is a single directing force. This need not be a large staff. At the giant IBM Corporation only one engineer is engaged full time in the conversion program (although obviously all supervisors, plant engineers, and machine operators are involved in the program to some extent). The

executive engineer in charge of the metric planning activity of the even larger General Motors Corporation has one assistant and a secretary.

(5) While central coordination is necessary, it has been found that problems connected with conversion get worked out more quickly and effectively on lower levels, where they actually occur, and that, therefore, it is wise to give supervisors considerable latitude in solving them. The supervisors should understand that whatever innovations they make should be reported to the coordinator, however. This system takes full advantage of what is usually called American ingenuity without sacrificing central control.

(6) Retraining of employees should be kept as simple as possible. Only those workers who have to be retrained should be included in the training progam, and they should be retrained only to the extent necessary. Thorough training programs for everyone in the plant, including file clerks, have proved to create unnecessary administrative, technical, and personnel problems. By creating the impression that a massive change in company procedure is taking place, such programs also tend to throw many employees into a panic. A low-keyed approach including only those employees who are going to be working directly with metric units, geared to the assuring premise that the metric system is easy to learn and more efficient than the inch system, has been found to make the transition much smoother.

In manufacturing plants where precision work is required, engineers and machine operators will already be used to working with decimals instead of fractions, even if they are unfamiliar with the metric system. For years industry has been stealing from the metric system by measuring precision work in tenths, hundredths, and thousandths of inches instead of in halves, fourths, eighths, sixteenths, thirty-seconds, and sixty-fourths. It will not be difficult for a lathe operator who is used to measuring diameters with a set of calipers calibrated to 1/5,000 of an inch to learn to use a set calibrated to 1/200 of a millimetre.

Companies have also found that workers should not be retrained until the new machines are available and they can begin to work on them as soon as they are retrained. A worker usually has a degree of enthusiasm to try out his new knowledge when he is first introduced to the metric system, and there is a consequent feeling of letdown if he discovers it is going to be some weeks, or even some months, before he will get a chance to use what he has just learned.

(7) The "rule of reason" should apply in deciding whether or not to convert a particular operation or machine. Conversion in any area should never be made "at all cost," but neither should it be avoided because of cost. The economic advantages and disadvantages of each individual area of conversion should be considered, and conversion should take place in that area only if there is an economic advantage. It should be kept in mind that there are only two reasons to convert: increased efficiency of operation, and increased market for the product. If one or the other doesn't justify the estimated direct cost, there is no point in making the conversion in that particular area. For example, there would be no economic advantage in changing the company's stationery from the customary 8½-by-11-inch size (which is 22.59 by 27.94 cm) to the nearest metric round numbers of 22.5 by 28 cm. Perhaps twenty-five or fifty years from now everything in the country will have been converted to metric standards, and 22.5 by 28 cm will be the stock stationery size, but for the moment there is no point in any company's undergoing a crash program in which *everything* is converted to metrics simply for the sake of conversion.

A curious sidelight on the metrication of all four of the big companies is that, while the decision to go metric, of course, came from top management, the inception of the idea came from the middle management echelon. In each case, middle management people had to sell top management on the idea.

Whether or not there is a lesson in this, I don't know; but if there is, perhaps it is that metric training in industry should start with company presidents, chairmen of the boards, and members of boards of directors.

The impact of metrication on the various packaging industries will probably not be as great as on the manufacturing industry. The packagers of food, soap powder, cosmetics, and other items that come in boxes, cans, and bottles do not contemplate having to make any major changes in machinery. In the beginning at least the only change will probably be the addition of metric units to the amounts in customary terms printed on the outside of packages. (Some packages already list contents in both customary and metric terms because they are sold both to the domestic market and to the foreign market.)

As an example of the problems faced by the packaging industry in converting entirely to metrics, consider General Mills, Inc., of Minneapolis, one of the largest packagers of food products. Most of the basic machinery in General Mills plants is long-lived; ordinarily, it is used for twenty to thirty years before replacement is necessary. Some of this machinery can be modified with new dials, scales, and meters that read in metric terms, but machines that can't are not likely to be scrapped by General Mills, because they are too expensive to replace. The company's tentative plans are to use present machinery until it wears out and to replace each piece with metric machinery when it is retired. Since the expected life of present machinery ranges upward to thirty years, total conversion may take that long.

This problem probably won't even be noticed by the average consumer. The demand for metric products is pretty well limited to those which might require repairs or replacement parts, and a box of cereal or a bag of flour can never require either.

Most packaging machinery, whether in the food industry or for nonedible goods, is adjustable enough so that when and if packagers do decide to change the amounts that they sell to even metric units, the quantities packaged can be either increased or decreased to the nearest even metric volume or weight. Initially, most companies will probably leave the boxes or cans in which their products are dispensed the same sizes and adjust the contents to the new amounts. For the most part, these adjustments should be small enough so that even slight *increases* will be possible, since package containers are almost never filled to their total capacity.

In instances where package sizes have to be changed, costs will be less for goods packaged in cardboard boxes than for those that come in cans or bottles. The Can Manufacturers Institute advises that if can sizes have to be changed, the least expensive change would be to change the height and leave present diameters as they are. Bottle manufacturers will have an even larger problem, because changes in bottle sizes will require extensive retooling.

When making any changes in package sizes, companies will necessarily have to take the following two factors into consideration.

(1) Many package sizes are designed for the convenience of the consumer. For example, a 10-ounce can of soup is supposed to contain servings for two people. Any substantial change in the contents of soup cans might be resented by consumers.

(2) Many states have laws regulating the package sizes of certain items offered for sale. For instance, most states permit the sale of such items as flour, cornmeal, and sugar only in weights of 2, 5, 10, 25, 50, and 100 pounds, and in multiples of 100 pounds. Butter must be sold in units of no less than ¼ pound. Milk, ice cream, and other dairy products can be sold only in multiples and subdivisions of the quart. This legislation would all have to be amended before packages containing these items could legally be changed to metric sizes. You,

therefore, should not anticipate buying milk and ice cream by the litre, even after the whole nation goes metric, until state legislatures get around to revising their packaging laws.

In 1974 the Wine Institute, a trade association of California wine producers, petitioned the U.S. Bureau of Alcohol, Tobacco and Firearms for amendment of the *Regulation on Labeling and Advertising of Wine* to require metric standards of fill and standard sizes and designs for wine bottles, both for domestically produced wine and wine imported in bottles. In June 1974 the Bureau held a public hearing attended by winemakers, representatives of numerous foreign and domestic trade organizations and of various foreign countries that import wine into the United States, and other interested persons. While there was some disagreement on the sizes of bottles and amounts of fill that should be permitted, no one voiced objection to metrication per se. The regulation was accordingly amended in December 1974 to require metrication by the wine industry and all wine importers by January 1, 1979.

The Wine Institute estimates the cost of conversion at over a quarter million dollars, mostly to the bottle manufacturers, who will be required to convert some 227 stock molds. According to Hugh Cook, Director of the Technical Division of the Wine Institute, the conversion is being made in the interest of the consumer, who will more easily be able to do comparison shopping when all wines are bottled to the same standard.

In other industries some packagers will have no problem because they are already on the metric system. It has been mentioned that the pharmaceutical industry quietly went metric some years ago. This does not apply merely to prescriptions. Many patent medicines and most vitamins are already bottled and boxed in metric units. It is probable that the rest will go that route much more quickly than other packaged goods, simply because the nonmetric packagers of medicines are already in the minority.

Chapter VIII

How Metrication Will Affect the Individual

When a writer encounters a friend or acquaintance he has not seen for some time, as soon as the customary pleasantries are out of the way, almost invariably he is asked, "What are you working on now?" While researching and writing this book, I was probably asked that question at least a hundred times.

The reaction to my saying that I was writing a book on the coming conversion to the metric system was varied. Some people showed only polite interest; some seemed surprised that I could find enough to say on such a subject to fill a book; still others seemed pleased at the idea, made comments indicating they thought conversion to metric was long overdue, and said they looked forward to reading the book when it came out. But an astonishing number of people were quite disturbed by the prospect of metrication.

One woman acquaintance said, "I hope it doesn't happen until after I'm dead and gone, because I know I could never learn all those foreign words."

People who are that apprehensive about metrication seem to be under the impression that on some specific date the use of the inch, the quart, and the pound will suddenly cease, and from that day on they will have to wrestle with the unfamiliar units of the metre, the litre, and the kilogram.

No such abrupt change is going to take place, of course. But on the other hand, we are not going to drift slowly into metrication either. We have been doing that for nearly 110 years, ever since Congress legalized the metric system in 1866, and at that same rate of drift we could expect complete metrication by perhaps 2066.

What is contemplated under the federal metrication program is not a massive and immediate change to the metric system, but merely a massive educational program designed to make the general public familiar with the metric system and to get people to accept it. The changeover will in no sense be sudden, but it is hoped that once conversion is started, it will move along at a fairly rapid clip. No one has even suggested that metrication can be accomplished in less than several years, and it is accepted as a matter of course that in a few areas there will never be conversion.

The same rule of reason will apply in these areas that industry uses in deciding if a specific change to metric is justified. These special areas may be roughly divided into two categories: (1) traditional matters in which the customary system of measurement is such an integral part that there would almost certainly be widespread resistance to change and (2) areas where it would be economic idiocy to attempt metrication.

Football is an example in the former category. The length of a football field will no doubt remain 100 yards instead of changing to 91.44 metres, and you are unlikely ever to hear a sports announcer say anything on the air such as, "Third down, one and eight tenths metres to go." Baseball diamonds and basketball courts will probably continue to be laid out in customary units also, at least in the foreseeable future. In track and field and swimming, however, distances that haven't already been metricized probably will be rather quickly, because Olympic measurements in these sports are all in metric terms. The Olympic-size pool is now common for swimming competition, and the 100-metre dash has in many places replaced the 100-yard dash. Broad jumps, high jumps, pole vaults, and distance throws are already measured in metric terms in many schools in order to make it easier to prepare for Olympic competition, and there would probably not be a great deal of opposition to complete metrication in track-and-field competition.

In the second category a good example are the manufacturers of oil-well equipment. Most of the well heads, fittings, tools and casing hangers used in oil fields throughout the world are manufactured in the United States. The few foreign manufacturers, even though all are in metric countries, fabricate them to customary standards in order to compete. It is unlikely this industry will go metric in the foreseeable future, simply because there is no competitive pressure to convert. Under the rule of reason it would be silly to go to the enormous expense of conversion when there is no economic advantage.

Not quite in this category, but in the category where the rule of reason shows no economic justification for conversion is an area touched on in Chapter VII. It is extremely unlikely that property records will be rewritten in metric terms at least within the next century.

The thrust of the federally directed metrication program will be to disrupt the everyday life of the average citizen as little as possible during the changeover, but to convert in those areas that count the most as quickly as is feasible without too much disruption. Areas that are considered to count most are those where metrication will either make computations easier or transactions between individuals simpler. Obviously, there would be neither of these advantages if football measurements were converted to metric, so no effort will be made to force such conversions as that on the general public.

It is conceivable that eventually, when the metric system has become the ordinary language of measurement to such a degree that most people have difficulty visualizing distances in yards, football fans will demand a change. If and when that ever happens, the field will probably be lengthened to 100 metres instead of merely being rounded off to 90 metres, which would leave it at approximately its present length, so that the same ten divisions can be maintained between goal lines. But this is not a prospect for sports fans to worry about at the moment, because if it ever happens, it is not likely to occur before the year 2000 at least.

When engineers speak of conversion to metrics, they use the terms *soft conversions* and *hard conversions*. Soft conversions are merely changes in terminology. Units are converted from the customary to the metric system without changes in the lengths, volumes, or masses measured. If and when the news media reports that a new Miss America has measurements of 91–63–89, that will be a soft conversion. A hard conversion means an actual change in size. If you are a manufacturer whose product requires ½-inch-diameter bolts, and you start calling them 12.7 mm bolts, you have made a soft conversion. If you switch to using one of the closest equivalent metric bolts of 12 mm or 14 mm diameter, it is a hard conversion. Similarly, if the manufacturer of the bolts you use merely converts their dimensions to metric terminology, it is a soft conversion. But if he retools in order to turn out bolts in standard metric sizes, he has made a hard conversion.

The first conversions the average consumer will encounter are likely to be soft conversions. Suddenly you will find yourself buying 454 grams of hamburger instead of 1 pound. Since that is an awkward amount to deal with, your butcher will probably quite soon change to selling hamburger in amounts of 500 grams and multiples of 500 grams. When he does that, it becomes a hard conversion. You will then start paying a little more for a package of hamburger than you previously did, because you will be buying slightly over 1½ ounces more hamburger than you are used to buying.

This principle will also apply to other food items sold in bulk. In the vegetable department of your favorite supermarket, you will discover that the old ounce-pound scales have been replaced by metric scales. Instead of 3 pounds, the scale will register 1½ kilograms when you place the amount of onions you usually buy on it. (Actually, this will be 40 grams more than 3 pounds, but that is as near as you can get to an equivalent round metric figure.) When you weigh what looks to you like about 10 pounds of potatoes, the scale will register slightly over 4½ kilograms (about 36 grams more than 10 pounds).

Probably, instead of weighing out such amounts as this, you will soon get in the habit of buying all but the more expensive bulk vegetables in round numbers of kilograms, which in some cases will mean buying a little less than you are accustomed to buying at present and in some cases, buying a little more. If you are accustomed to buying 3 pounds of onions at a time, you will probably start buying 1 kilogram at a time, which is about 2⅕ pounds. If you usually buy potatoes in 10-pound lots, you will probably start buying them in 5-kilogram lots, which is about 11 pounds. In any event, within a very short time you should be able to visualize just how much of any item you require in terms of grams and kilograms as easily as you can now visualize your requirements in terms of ounces and pounds.

Selling bulk foods in metric units involves such a simple soft conversion that it will probably take place very soon after the announced date for the beginning of the change to a metric America. The appearance of prepackaged foods such as bacon, sausage, hot dogs, and cheese in metric amounts will probably be slower. As mentioned in the previous chapter, the first such changes are likely to be simply the addition of metric equivalents to the customary weights given on packages. A package of bacon will then read: *1 lb.: 454 g.*

When and if packagers change the size of packages to round out their contents to even numbers of grams, this dual listing of weights will probably be stopped, and thereafter you will buy bacon, sausage, hot dogs, and cheese in units of 500 grams or in kilograms. This change will involve a hard conversion for the food packager, since there is a change in quantity, not merely a change in terminology.

The term *hardware conversion* refers to the replacement of a customary-system tool or machine by a metric one. When the Miss America Contest judges at Atlantic City discard their inch-foot tape measure in favor of a metric tape measure, that will be a hardware conversion. When your local radio station replaces its Fahrenheit outdoor thermometer with a Celsius thermometer so that temperatures can be broadcast in metric terms, that also will be a hardware conversion.

The term originally was coined to describe tool and machine changes in industry, but as indicated by the above examples, it has come to include changes from customary-system devices to metric ones anywhere. You will find that over a period of time you are making numerous hardware conversions in your home, because many household devices involve measurement.

The simplest hardware conversion that will take place in the home will be when you throw away your wooden yard-stick and buy a metre stick. When your bathroom scale wears out and you buy another one that registers your weight in kilograms and hectograms (it is not likely to weigh any finer than one tenth of a kilogram), you will have made another hardware conversion. A new kitchen stove with a Celsius temperature scale on the oven would constitute a third hardware conversion.

You can see from these examples that except for very simple and inexpensive hardware, most conversion in the home for many years is going to be soft. People will not discard their bathroom scales until the scales no longer work. Cooking stoves sometimes remain in use for twenty to thirty years. Thermometers, as mentioned in a previous chapter, are not going to be thrown away so long as they continue to work. So even long after the metric system has come into common use, millions of homes will still contain unconverted hardware items.

Once metric has become the common language, the use of such hardware items will require constant soft conversion of customary units to metric. No doubt enterprising companies will meet this need by marketing small, convenient weight-conversion tables to hang on bathroom walls and even smaller decals containing temperature-conversion scales to paste on kitchen stoves. Some stove manufacturers may even market Celsius oven-temperature dials, along with instructions for their installation.

The whole problem of home hardware conversion will eventually be solved simply by the passage of time, of course. Once the nation has definitely gone metric, newly married couples will start out with all-metric appliances. Even if a couple can afford only a secondhand stove, the bride is likely to insist on one with a Celsius oven gauge. This will mean that stoves with Fahrenheit gauges will have little if any re-sale value, and they will probably simply be scrapped when their present owners are through with them.

The biggest day-to-day conversion problem confronting the average housewife will involve cooking. In the beginning there will be little problem because her standard cookbooks will all be written to customary measurements, and she will continue to use her old measuring devices, such as table-spoons, teaspoons, and measuring cups graduated in ounces.

However, many housewives like to experiment with new recipes, and are constantly cutting them from newspapers and magazines or copying them down from TV cooking shows. They also are likely to acquire new cookbooks periodically, either as gifts or by personal purchase.

At first it is probable that recipes gathered from the media and those published in new cookbooks will give both metric measurements and customary measurements. But under the Think Metric concept, the public education program spon-sored by the federal government is almost certain to include strong urging against this practice. The experience in England proved pretty conclusively that when there is an attempt to ease the transition to the metric system by publicly condoning the continued use of the customary system alongside the metric system until the public has learned the latter, little attempt is made to learn metrics, and the use of the customary system continues indefinitely. The experience in Australia proved equally conclusively that when the use of the customary system is actively discouraged, people very quickly learn and prefer to use the metric system.

It is likely that the news media and the publishing industry will both go along with the federal government's request to give new recipes in metric terms only. This is going to force the housewife to use metric units, whether she likes it or not.

Her first reaction will probably be simply to convert recipes from metric to customary terms and go right on using the same old measuring equipment. If the Think Metric program becomes really successful, however, housewives will find it more and more difficult to find conversion tables enabling them to do this. Just as I have included only tables for converting from the customary system to metric in this book, with the deliberate purpose of forcing the reader to Think Metric, authors of future cookbooks are likely to adopt the same tactic with the same motive. To paraphrase an expression that became a cliché during a recent televised senatorial investigation, when that point in time is reached, Mrs. Housewife is going to have to scrap her customary measuring devices and go out and buy new ones.

This, of course, will be a hardware conversion, but hardly an expensive one. Because very few kitchens are equipped with weight scales, recipes are always written in terms of volumes, with the exception of ingredients that are assumed to have been preweighed. Thus if a recipe calls for 500 grams of ground round steak, it is assumed the housewife will have her butcher weigh her out that amount. If the recipe calls for 125 grams of butter, it is assumed she knows enough merely to unwrap one of the four sticks of butter in a 500-gram package. In general, when cookbooks are written, it is taken for granted that there will be no food scale in the kitchen.

Since the housewife does not have to worry about converting as expensive an item as a weight scale, hardware conversion in the kitchen should be relatively simple. The only measuring tools in the average kitchen are measuring cups and spoons, which are inexpensive items.

There is no reason to believe that metric cups and measuring spoons will sell for much more when they appear. A customary measuring cup holds up to 8 ounces. The American National Standards Institute has recommended that the metric 1-cup measure be based on the litre, and its capacity be 250 millilitres, or 0.25 litre. This is only slightly more than the present 8-ounce cup. The customary measuring-spoon set includes a tablespoon measure, a teaspoon measure, a half-teaspoon measure, and a quarter-teaspoon measure. A metric set of spoons for dry measure includes five instead of four: 1, 2, 3, 5, and 10 cm^3. By using these in combination, you can make any measure from 1 to 20 cubic centimetres (the equivalent of 1 tablespoon plus 1 teaspoon).

Another type of home hardware that is going to require constant soft conversion is baking pans. This is going to be a very minor problem, though. Standard round pie and cake tins measure 9 inches in diameter. Square cake tins measure 8 inches across. Nine inches is 22.86 centimetres; 8 inches is 20.32 centimetres. The standard metric round pie and cake tins measure 21 centimetres in diameter; the standard square cake tin is 21 centimetres across. In both cases the difference between customary-measurement tins and metric tins is too small to make any difference in a recipe, so housewives can continue to use their present baking pans and merely remember that when a recipe calls for a 21-centimetre round pan, it means the standard 9-inch pan, and when it calls for a 21-centimetre square cake pan, it means the standard 8-inch pan.

The first soft conversions you will encounter outside of the home, other than at the supermarket, are likely to be in road and highway signs. Insofar as speed limits are concerned, here again adjustments will be made to the nearest round figure. Twenty-five miles per hour happens to convert very closely to 40 kilometres per hour, but 30 miles per hour converts to about 48 kilometres per hour. The latter will probably be raised to 50 kilometres per hour. Fifty miles converts very closely to 80 kilometres.

Distances, of course, will be converted exactly, to the nearest tenth of a kilometre, just as they are presently given to the nearest tenth of a mile.

Until all automobiles can be equipped with metric speedometers, distances and speed limits will have to be posted in both metric and customary units. There would be too much chance of error if motorists had to make constant conversions from metric to customary in their heads, not to mention the increased chance of accidents because of their mental preoccupation. This is one compromise with the Think Metric approach that the federal coordinators are willing to accept for safety reasons.

Other soft conversions that will be encountered early in the changeover program will be the sale of items in metres that you are used to buying by the yard. This will include cloth, draw cords for drapes, wire, and a host of other items sold by length.

The pumping of gasoline by the litre instead of by the gallon will probably lag behind the installation of new road signs, because this involves a hardware conversion. The gallon meters on pumps will have to be replaced with meters that record in litres. However, the pumps won't have to be replaced, so this conversion should not be too difficult.

If the federally directed conversion program develops as the Department of Commerce hopes, the process of conversion should be relatively painless for the average individual.

Chapter IX

The National Bureau of Standards

As one of the forty-two signatories to the Treaty of the Metre, the United States recognizes the International Organization of Weights and Measures, founded by that treaty, as the supreme international authority on the subject. The governing body of this organization is the General Conference on Weights and Measures, which has already been described. Below this is the International Committee on Weights and Measures, which carries on the organization's work between meetings of the General Conference. Below it is the International Bureau of Weights and Measures, which is responsible for the maintenance of international standards. Standards are maintained at the bureau's laboratory in Sèvres, France.

Beneath the International Committee there are also a number of consultative committees. These committees make recommendations concerning their specialized areas to the International Committee, which in turn passes them on to the General Conference at one of its periodic meetings. These committees have such titles as the Consultative Committee on Units and the Consultative Committee on Weights.

Most countries that are party to the Treaty of the Metre also have national bodies responsible for the maintenance of standards in their own countries. France, for instance, despite being the location of the International Organization of Weights and Measures, also has the *Laboratoire d'Essais*, which is responsible for the supervision of weights and measures in

France only. In Italy a similar body is called the *Servìzio Centrale Mètrico.*

In the United States the body responsible for the supervision of weights and measures and for the maintenance of standards is the National Bureau of Standards.

The National Bureau of Standards came into existence because of an acute need. At the turn of the century, standardization, or rather the lack of standardization, was becoming an increasing problem for both science and industry.

As an example, consider the electric light bulb. When you buy a new electric light bulb, you expect it as a matter of course to fit any lamp or wall socket you screw it into. But when the incandescent lamp first came on the market, 175 different sizes of the screw-in part were soon available.

By 1900 many manufacturers were working out industry-wide agreements limiting the manufacture of certain products to a specific range of sizes. But they ran into the problem that measuring equipment itself was not standard enough to give these agreements much meaning.

At its annual meeting in 1900 the National Academy of Sciences declared:

The facilities at the disposal of the Government and of the scientific men of the country for the standardization of apparatus used in scientific research and in the arts are now either absent or entirely inadequate, so that it becomes necessary in most instances to send such apparatus abroad for comparison.

The Academy of Sciences was referring to such ridiculous things as the United States Navy's having to send its navigational instruments to Germany for calibration because there was no testing laboratory in the United States capable of doing this work. Europe had a number of such laboratories, where the accuracy of any type of measuring instrument from a thermometer to a weight scale could be certified. Scientific laboratories in America generally used only instruments so certified, with the result that American instrument manufacturers found themselves unable to sell to the scientists of their own country.

The Office of Weights and Measures, mentioned in Chapter I as a part of the Treasury Department, had become largely nonfunctioning by 1900. Operating on a budget of only $10,000 a year, it did little more than act as custodian of the copies of the kilogram and metre prototypes sent from France ten years before. The metre bars provided by the International Bureau of Weights and Measures, incidentally, merely had the official length of the metre marked on them. The nations receiving them were expected to calibrate them into subdivisions. This was a highly technical job, requiring microscopic precision of workmanship, and the Office of Weights and Measures was never able to afford to have it done.

On March 3, 1901, Congress abolished the Office of Weights and Measures and established the National Bureau of Standards. In 1903, when Congress established the new Department of Commerce and Labor, the bureau was placed under it. In 1913, when a separate Department of Labor was created, the bureau remained under the Department of Commerce, where it still is.

Under the original act, the bureau was given custody of basic standards and was empowered to test and certify instruments submitted to it for a fee, a service it still performs. It was also empowered to construct new standards when necessary. And it was authorized to conduct scientific research in a number of areas.

Over the years the bureau has grown in both size and responsibility, as additional legislation has given it new functions. At the present time it is divided into four subdivisions, plus an office of information.

The Institute for Basic Standards

The Institute for Basic Standards is responsible for maintaining standards of measurement and for coordinating these standards with those of other nations. It also furnishes services designed to make accurate and uniform the measurements used throughout the country by science, industry, and commerce. There are twelve divisions within the agency, which are separately concerned with: (1) applied mathematics, (2) astrophysics, (3) atomic physics, (4) chemistry, (5) electricity, (6) heat, (7) mechanics, (8) metrology, (9) physics, (10) radiation, (11) radio engineering, and (12) radio physics.

Let's consider one example of how the institute's services "make accurate and uniform the measurements used throughout the country." The electric-distributing companies employ a device called the rotary standard watt-hour meter. This is similar to the electric meter in people's homes, except that it is kept in the laboratory of the electric company and is used as a standard to check the accuracy of home meters. The instruments used to measure the accuracy of home meters are checked against this central meter before inspectors and technicians use them in the field. The importance of these central meters' being accurate can hardly be overestimated when you consider that there are something like sixty million home electric meters in the United States, all recording the amount of electricity being used, and thus determining the monthly electric bill for each household. It is estimated that if they were all either slow or fast by 1 percent, customers would be undercharged or overcharged by a national total of more than a hundred million dollars a year.

In order to make sure that the basic meters against which it checks home meters are accurate, each electric company periodically sends its basic meter to the Institute for Basic Standards for tests. Such tests cost the company about seventy dollars each time.

The institute does not adjust these meters if they are found to be inaccurate, but merely reports the degree of inaccuracy. The company can either make its own adjustment or leave the instrument as it is and take the percentage of error into

The National Bureau of Standards

account when the inspectors and technicians check
measuring instruments against it.

Electric meters are only one of the many measurin;
vices the institute constantly checks for accuracy. In
cases the measuring devices industries and laboratories
in for checking are, like the electric meter, master instrun
After they have been certified by the National Bure;
Standards, all of the industry's or the laboratory's other ii
ments of similar nature are checked against it. In this
uniform standards are passed down from the bureau thr
out the nation.

The Institute for Materials Research

The Institute for Materials Research conducts researcl
assists industry in developing research programs of its
concerned with discovering improved methods of meas
the basic properties of materials used by industry, comn
educational institutions, and government agencies. It ¢
ops measurement standards for various materials and
vides advisory and research services to other governn
agencies. Its various divisions are oriented more towar
search concerned with measurement than toward the ;
testing of measurement equipment.

The concerns of its separate divisions include: (1) an
chemistry, (2) cryogenics, (3) inorganic materials, (4) n
lurgy, (5) polymers, and (6) reactor radiations. Each di\
is responsible for developing methods of measuring the
erties of materials within its own particular field.

Government bureaus, particularly scientific bureau;
hardly hotbeds of controversy under ordinary circumst;
but the type of work done by the Institute for Material
search has twice created national uproars.

The first goes clear back to 1928, when a consumer ;
cate, a forerunner of Ralph Nader, publicly charged that
federal government, whose officials profess solicitude fc
consumer, possesses a vast store of useful knowledge wh

refuses to make available to the general public because it would do damage to business."

He was referring to tests of certain products made by the National Bureau of Standards, the results of which he felt the general public was entitled to know, but which he claimed the bureau was deliberately suppressing because of political pressure from manufacturers. The implication was that results were being suppressed because the tests all showed the products were either inferior or useless.

The charge brought some public clamor for the bureau to release its test results, but the bureau refused. The fact was, and still is, that the bureau was never meant to be an agency for consumer protection but, as its name implies, is concerned with measurement standards. In that role it often releases the results of tests on specific materials, but never by brand names.

Many people are under the misapprehension that one of its functions is to test brand-name products, however, and the Institute for Materials Research receives thousands of letters every year from consumers asking for information about different products. Occasionally, a manufacturer even expects the institute to test his product so that, if the institute gives it a good mark, he can advertise that fact.

Such requests from both consumers and manufacturers receive the polite answer that the institute does not either approve or disapprove brand-name products.

The only exception it ever made to that rule brought on an even bigger clamor than the one in 1928. In 1951 the institute released the results of its tests on additives which manufacturers claimed would restore life to dead storage batteries. About thirty different additives had been tested. In conformity with bureau policy, none were mentioned by name, but the report was that all of them were useless.

Jess Ritchie, the owner of a small company in Oakland, California, that manufactured a battery additive called AD-X2, complained that this report reflected on his product.

Insisting that AD-X2 actually did restore dead batteries, he demanded a special test of his product.

When the bureau informed him that AD-X2 had been one of the additives tested, Ritchie rejoined that the test must have been improperly conducted and began a campaign to get the additive retested.

There is some difference of opinion among observers of subsequent events as to whether Jess Ritchie had powerful political influence or merely raised so much disturbance that he got himself noticed. Proponents of the first theory point out that twenty-eight different senators—including one named Richard Nixon—contacted the bureau on Mr. Ritchie's behalf, requesting that AD-X2 be given a new test. Proponents of the second theory insist that he was merely a small businessman without political influence who elicited the sympathy of all those legislators because he convinced them he was being discriminated against by an impersonal governmental bureaucracy. It was later revealed that Ritchie had urged his distributors all over the country to write to their senators about the matter, which gives some credence to the second point of view. But the peculiar behavior of Secretary of Commerce Sinclair Weeks tends to give just as much credence to the first.

About two years after the controversy first began, in 1953 (when former Senator Richard Nixon was now Vice President Nixon), National Bureau of Standards Director Allen Astin finally gave in to the pressure and had AD-X2 retested, with representatives of Ritchie's company present as observers. When it was announced that the test results were the same as before, Secretary of Commerce Weeks made the astonishing announcement that the Bureau of Standards' testing procedure was faulty, that he personally had used AD-X2 and knew that it worked, and demanded Director Astin's resignation.

He got the resignation, but he simultaneously got a hurricane of reaction. Allen Astin was a highly respected scientist and had been with the bureau for twenty years. Fifty key scientists in the bureau announced that they would resign if Astin was not reinstated. A few days later various division chiefs announced that they had conducted a secret poll of all divisions, and had turned up nearly four hundred more scientists and engineers in the bureau who were considering resignation if Astin was not reinstated. Newspapers printed editorials accusing Weeks of bringing blatant political pressure on an agency that should be above politics. *Newsweek,* the *Nation,* the *New Republic,* and *Science Newsletter* all published magazine articles critical of the Secretary of Commerce.

Sinclair Weeks decided to temporize by appointing a committee of scientists to investigate the whole matter and recommend action. When the committee found Astin faultless, he was not only reinstated, but Weeks publicly assured him that he had the job of director as long as he wished.

Jess Ritchie managed to get personal vindication also, though. He had an independent laboratory run a new set of tests on AD-X2, and its report was that the additive *did* restore dead batteries. Meantime the national publicity generated by the case brought him a flood of orders for his additive from all over the country. Oddly, the initial bad report by the bureau actually increased his sales more than it would have if it had been a good one—which probably would have gone largely unnoticed.

One function of the Institute for Materials Research is to test products for agencies of the federal government. In this capacity it has probably at some time or other tested every type of product manufactured, except food and drugs, which are the province of the Food and Drug Administration. This does not mean that it has tested every *product* manufactured, but only some of each type.

In making such tests for federal agencies, it makes no attempt to rate the qualities of products, but merely lists their specifications. This is because governmental agencies buy most things by specifications instead of by quality. Purchases are generally made in such huge quantities that it would be economically unsound always to buy the very best product, when as often as not a product of lesser quality will serve the purpose. Therefore the agency in the market for a product figures out what the product's minimum qualifications have to be in order to fulfill its purpose, and lists these as specifications.

For instance, the United States Navy uses a tremendous amount of gray paint on its warships. Its specifications for paint will include a certain minimum lead content, a certain pigment content, the requirement that the paint meet certain standards of resistance to wear, and perhaps a half dozen other specifications. These won't necessarily be the specifications for a top-grade maritime paint, but they will be the ones the Navy, through experience, has found adequate for its purposes.

When these specifications are published, various paint companies submit bids to the Navy Procurement Office, along with samples of paints mixed to the specifications required. The Navy then requests the Institute for Materials Research to test these samples *to see if they meet specifications.*

This is not a request for a comparison of the qualities of the various samples, but only for one of two ratings on each: either it meets specifications, or it doesn't. And that is all the test results report. None of the companies can later advertise that their paints are "government-approved," because neither approval nor disapproval is involved in this type of report.

When all the reports are in, the Navy customarily awards the contract to the company making the lowest bid among those whose paint met the minimum specifications.

The institute's most important function, insofar as assisting industry is concerned, is its research designed to find new ways to measure the properties of materials. These research results are available to any manufacturer, business, or educational institution, since they involve merely procedures for measuring the qualities of various materials and make no ratings of the materials themselves.

The general public's memory of the AD-X2 affair has apparently faded, because the institute frequently receives suggestions that public release of all the information about products in its files would benefit the general public. The routine polite reply is that the institute is not an agency for consumer protection.

The Institute for Applied Technology

The Institute for Applied Technology develops methods of measuring and evaluating the performances of machines, appliances, and other technological products and services. It cooperates with public and private organizations in the development of technological standards and test methods, and it advises and does research for federal, state, and local governmental agencies.

Its divisions include (1) the Building Research Division, (2) the Electronic Instrumentation Division, (3) the Technical Analysis Division, and (4) the Textile and Apparel Technology Center.

Typical of this institute's activities was the development of a surge generator capable of developing 11 million volts in order to create artificial lightning. The device is used to measure the effectiveness of porcelain insulators used on high-tension wires, which are often targets of lightning.

This institute also maintains the Clearinghouse for Federal Scientific and Technical Information, a vast file of technical reports on virtually every scientific subject.

The Center for Computer Sciences and Technology

The Center for Computer Sciences and Technology conducts research and provides technical services to help government agencies select and learn to use automatic data-processing equipment. For a fee, it provides a computer-programming service and acts as a consultant in the development of computer systems. It is also reponsible for developing and monitoring voluntary standards in the computer field.

Since 1966 the National Bureau of Standards has been housed in a $120-million facility at Gaithersburg, Maryland, near Washington, D.C. It consists of an administration building, several general-purpose laboratories, and thirteen specialized laboratories. In the planning stage are an industrial building, a sound laboratory, a concrete-materials building, a fluid-mechanics laboratory, and a nonmagnetic laboratory.

When an industry, business, educational institution, or individual brings a measurement problem to the National Bureau of Standards, the problem may be presented in terms of customary units, or it may be in metric terms. In making its tests and in rendering its report, the Bureau uses whichever measurement system was used in presenting the problem. But for all scientific work within the bureau itself, the International System of Units is used.

Hopefully, that will be the universal measurement system in the United States before long.

Appendix A

The sign · means multiplied by.
The sign / means divided by.

SI Base Units

Quantity	Unit	Symbol
length	metre	m
mass	kilogram	kg
time	second	s
electric current	ampere	A
thermodynamic temperature	kelvin *	K
amount of substance	mole	mol
luminous intensity	candela	cd

* SI also accepts the degree Celsius (C) as equivalent to the kelvin.

Length Units

Unit	Value	Symbol
kilometre	1 000 metres	km
metre	base unit	m
decimetre	0.1 metre	dm
centimetre	0.01 metre	cm
millimetre	0.001 metre	mm

Mass Units

Unit	Value	Symbol
kilogram	base unit	kg
gram	0.001 kg	g
decigram	0.1 g	dg
centigram	0.01 g	cg
milligram	0.001 g	mg

TIME UNITS

Unit	Value	Symbol
second	base unit	s
minute	60 s	min
hour	3 600 s	h
day	86 400 s	d

VOLUME UNITS

Unit	Value	Symbol
cubic metre	derived unit	m^3
cubic decimetre	0.001 cubic metre	dm^3
cubic centimetre	0.000 001 cubic metre	cm^3
cubic millimetre	0.000 000 001 cubic metre	mm^3

SOME SI DERIVED UNITS

Quantity	Unit	Symbol
acceleration	metre per second squared	m/s^2
activity (radioactive)	1 per second	s^{-1}
area	square metre	m^2
capacitance	farad	F
concentration of amount of substance	mole per cubic metre	mol/m^3
conductance	siemens	S
current density	ampere per square metre	A/m^2
density, mass density	kilogram per cubic metre	kg/m^3
dynamic viscosity	pascal second	Pa·s
electrical potential, potential difference, electromotive force	volt	V
electric charge density	coulomb per cubic metre	C/m^3

SOME SI DERIVED UNITS (*Cont'd*)

Quantity	Unit	Symbol
electric field strength	volt per metre	V/m
electric flux density	coulomb per square metre	C/m^2
electric resistance	ohm	Ω
energy density	joule per cubic metre	J/m^3
energy, work, quantity of heat	joule	J
force	newton	N
frequency	hertz	Hz
heat capacity, entropy	joule per kelvin	J/K
heat flux density, irradiance	watt per square metre	W/m^2
illuminance	lux	lx
inductance	henry	H
luminance	candela per square metre	cd/m^2
luminous flux	lumen	lm
magnetic field strength	ampere per metre	A/m
magnetic flux	weber	Wb
magnetic flux density	tesla	T
molar energy	joule per mole	J/mol
molar entropy, molar heat capacity	joule per mole kelvin	J/(mol·K)
moment of force	metre newton	N·m
permeability	henry per metre	H/m
permittivity	farad per metre	F/m
power, radiant flux	watt	W
pressure	pascal	Pa
quantity of electricity, electric charge	coulomb	C
specific energy	joule per kilogram	J/kg
specific heat capacity, specific entropy	joule per kilogram kelvin	J/(kg·K)
specific volume	cubic metre per kilogram	m^3/kg
speed, velocity	metre per second	m/s
surface tension	newton per metre	N/m
thermal conductivity	watt per metre kelvin	W/(m·K)
volume	cubic metre	m^3
wave number	1 per metre	m^{-1}

SI Supplementary Units

Quantity	Unit	Symbol
plane angle	radian	rad
solid angle	steradian	sr

Some SI Derived Units from Supplementary Units

Quantity	Unit	Symbol
angular velocity	radian per second	rad/s
angular acceleration	radian per second squared	rad/s^2
radiant intensity	watt per steradian	W/sr
radiance	watt per square metre steradian	W·m^{-2}·sr^{-1}

Accepted Units from Outside of SI

Quantity	Unit	Value in SI Units	Symbol
time	minute	60 s	min
time	hour	3 600 s	h
time	day	86 400 s	d
angle	degree	$(\pi/180)$ rad	°
angle	minute	$(\pi/10\ 800)$ rad	′
angle	second	$(\pi/648\ 000)$ rad	″
volume	litre	dm^3	l
mass	tonne	10^{3}kg	t

Prefixes for Multiples

Multiplying Factor	Prefix	Symbol
10	deka	da
100	hecto	h
1 000	kilo	k
1 000 000	mega	M
1 000 000 000	giga	G
1 000 000 000 000	tera	T

Multiplying Factor	Prefix	Symbol
0.1	deci	d
0.01	centi	c
0.001	milli	m
0.000 001	micro	μ
0.000 000 001	nano	n
0.000 000 000 001	pico	p

TEMPORARILY ACCEPTED UNITS FROM OUTSIDE OF SI

Quantity	Unit	Value in SI Units	Symbol
acceleration	gal	1 cm/s^2	Gal
activity (radioactive)	curie	$3.7 \cdot 10^{10} \text{ s}^{-1}$	Ci
area	barn	10^{-28}m^2	b
area	are	10 m^2	a
area	hectare	100 m^2	ha
length	ångstrom	0.1 nm	Å
length	nautical mile	1 852 m	none
pressure	bar	100 000 Pa	bar
pressure	standard atmosphere	101 325 Pa	atm
radiation	röntgen	$2.58 \cdot 10^{-4} \text{C/kg}$	R
specific energy	rad	10^{-2}J/kg	rad *
speed	knot	$(1\ 852/3\ 600)$ m/s	none

* When there is risk of confusion with the symbol for radian, rd is used as the symbol for rad.

Appendix B

Conversion Factors

inches × 2.54 = cm
feet × 0.304 8 = m
miles × 1.609 = km
ounces of mass × 28.35 = g

pounds × 0.453 6 = kg
liquid ounces × 29.5729 = cm³
liquid quarts × 0.946 = l
gallons × 3.785 = l

Conversion Tables

INCH FRACTIONS TO METRIC

Inches	Millimetres	Inches	Millimetres
1/64	0.396 9	7/32	5.556 3
1/32	0.793 8	15/64	5.953 1
3/64	1.190 6	1/4	6.35
1/16	1.587 5	17/64	6.746 9
5/64	1.984 4	9/32	7.143 8
3/32	2.381 3	19/64	7.540 6
7/64	2.778 1	5/16	7.937 5
1/8	3.175	21/64	8.334 4
9/64	3.571 9	11/32	8.731 3
5/32	3.968 8	23/64	9.128 1
11/64	4.365 6	3/8	9.525
3/16	4.762 5	25/64	9.921 9
13/64	5.159 4		
13/32	1.031 9	23/32	1.825 6
27/64	1.071 6	47/64	1.865 3
7/16	1.111 3	3/4	1.905
29/64	1.150 9	49/64	1.944 7
15/32	1.190 6	25/32	1.984 4
31/64	1.230 3	51/64	2.024 1
1/2	1.27	13/16	2.063 8
33/64	1.309 7	53/64	2.103 4
17/32	1.349 4	27/32	2.143 1
35/64	1.389 1	55/64	2.182 8
9/16	1.428 8	7/8	2.222 3
37/64	1.468 4	57/64	2.262 2

Inch Fractions to Metric (Cont'd)

19/32	1.508 1	29/32	2.301 9
39/64	1.547 8	59/64	2.341 6
5/8	1.587 5	15/16	2.381 3
41/64	1.627 2	61/64	2.420 9
21/32	1.666 9	31/32	2.460 6
43/64	1.706 6	63/64	2.500 3
11/16	1.746 3	1	2.54
45/64	1.785 9		

Inch Decimals to Metric

Inches	Millimetres	Inches	Millimetres
0.001	0.025 4	0.06	1.524
0.002	0.050 8	0.07	1.778
0.003	0.076 2	0.08	2.032
0.004	0.101 6	0.09	2.286
0.005	0.127	0.1	2.54
0.006	0.152 4	0.2	5.08
0.007	0.177 8	0.3	7.62
0.008	0.203 2	0.4	10.16
0.009	0.228 6	0.5	12.7
0.01	0.254	0.6	15.24
0.02	0.508	0.7	17.78
0.03	0.762	0.8	20.32
0.04	1.016	0.9	22.86
0.05	1.27	1	25.4

Whole Inches to Metric

Inches	Centimetres	Inches	Centimetres
1	2.54	7	17.78
2	5.08	8	20.32
3	7.62	9	22.86
4	10.16	10	25.4
5	12.7	11	27.94
6	15.24	12	30.48

FEET TO METRIC

Feet	Centimetres	Feet	Metres
1	30.48	4	1.219 2
2	60.96	5	1.524
3	91.44	6	1.828 8
		7	2.133 6
		8	2.438 4
		9	2.743 2
		10	3.048

MILES TO METRIC

Miles	Kilometres	Miles	Kilometres
1	1.609	6	9.656
2	3.219	7	11.265
3	4.828	8	12.875
4	6.437	9	14.484
5	8.047	10	16.09

OUNCES (WEIGHT) TO METRIC

Ounces	Kilograms	Ounces	Kilograms
1	0.028 35	9	0.255 15
2	0.056 7	10	0.283 5
3	0.085 05	11	0.311 85
4	0.113 4	12	0.340 2
5	0.141 75	13	0.368 55
6	0.170 1	14	0.396 9
7	0.198 45	15	0.425 25
8	0.226 8	16	0.453 6

POUNDS TO METRIC

Pounds	Kilograms	Pounds	Kilograms
1	0.453 6	6	2.721 6
2	0.907 2	7	3.175 2
3	1.360 8	8	3.628 8
4	1.814 4	9	4.082 4
5	2.268	10	4.536

Ounces (Liquid) to Metric

Ounces	Litres	Ounces	Litres
1	0.029 572 9	17	0.502 739 3
2	0.059 145 8	18	0.532 312 2
3	0.088 718 7	19	0.561 885 1
4	0.118 291 6	20	0.591 458
5	0.147 864 5	21	0.621 030 9
6	0.177 437 4	22	0.650 603 8
7	0.207 010 3	23	0.680 176 7
8	0.236 583 2	24	0.709 749 6
9	0.266 156 1	25	0.739 322 5
10	0.295 729	26	0.768 895 4
11	0.325 301 9	27	0.798 468 3
12	0.354 874 8	28	0.828 041 2
13	0.384 447 7	29	0.857 614 1
14	0.414 020 6	30	0.887 187
15	0.443 593 5	31	0.916 759 9
16	0.473 166 4	32	0.946 332 8

Quarts (Liquid) to Metric

Quarts	Litres	Quarts	Litres
1	0.946 332 8	3	2.838 998 4
2	1.892 665 6	4	3.785 331 2

Gallons to Metric

Gallons	Litres	Gallons	Litres
1	3.785 331 2	6	22.711 987 2
2	7.570 662 4	7	26.497 318 4
3	11.355 993 6	8	30.282 649 6
4	15.141 324 8	9	34.067 980 8
5	18.926 656	10	37.853 312

Suggested Further Reading

Bendick, Jeanne, *How Much and How Many*. New York: McGraw-Hill, 1947.

Berriman, Algernon E., *Historical Metrology*. New York: E. P. Dutton, 1953.

Burton, William K. (ed.), *Measuring Systems and Standards Organizations*. New York: American National Standards Institute, undated, c. 1970.

Chiswell, B., and E. C. Grigg, *S.I. Units*. Sydney, Australia: John Wiley & Sons Australasia Pty. Ltd., 1971, and John Wiley, New York.

Perry, John, *The Story of Standards*. New York: Funk & Wagnalls, 1955.

United States Department of Commerce, *A History of the National Bureau of Standards*. Washington, D.C.: United States Government Printing Office, 1966.

United States Department of Commerce, *The International System of Units (SI)*. National Bureau of Standards Special Publication 330, 1972 edition. Edited by Chester H. Page and Paul Vigoureux. Washington, D.C.: United States Government Printing Office, April 1972.

United States Department of Commerce, Thirteen United States Metric Substudy Reports. Washington, D.C.: United States Government Printing Office. Listed below by National Bureau of Standards special publication number and dates issued:

NBS SP345-1: *International Standards*, December 1970.

NBS SP345-2: *Federal Government: Civilian Agencies*, July 1971.

NBS SP345-3: *Commercial Weights and Measures*, July 1971.

NBS SP345-4: *The Manufacturing Industry*, July 1971.

NBS SP345-5: *Nonmanufacturing Business*, August 1971.

NBS SP345–6: *Education,* July 1971.
NBS SP345–7: *The Consumer,* July 1971.
NBS SP345–8: *International Trade,* August 1971.
NBS SP345–9: *Department of Defense,* July 1971.
NBS SP345–10: *A History of the Metric System Controversy in the United States,* August 1971.
NBS SP345–11: *Engineering Standards,* July 1971.
NBS SP345–12: *Testimony of Nationally Representative Groups,* July 1971.
NBS SP345–13: *A Metric America: A Decision Whose Time Has Come,* August 1971. (This is the comprehensive report on the United States Metric Study and summarizes the findings of the above twelve reports.)

Index